T0292950

Exploring Complex Survey Data Analysis Using R

Surveys are powerful tools for gathering information, uncovering insights, and facilitating decision-making. However, to ensure the accurate interpretation of results, they require specific analysis methods. In this book, readers embark on an in-depth journey into conducting complex survey analysis with the {srvyr} package and tidyverse family of functions from the R programming language. Intended for intermediate R users familiar with the basics of the tidyverse, this book gives readers a deeper understanding of applying appropriate survey analysis techniques using {srvyr}, {survey}, and other related packages. With practical walkthroughs featuring real-world datasets, such as the American National Election Studies and Residential Energy Consumption Survey, readers will develop the skills necessary to perform impactful survey analysis on survey data collected through a randomized sample design. Additionally, this book teaches readers how to interpret and communicate results of survey data effectively.

Key Features:
* Uses the {srvyr} package and tidyverse family of packages.
* Grants a conceptual understanding of the statistical methods that the functions apply to.
* Includes practical walkthroughs using publicly available survey data.
* Provides the reader with the tools for interpreting, visualizing, and presenting results.

Stephanie A. Zimmer, PhD, is a senior survey statistician with 10 years experience in survey sampling and design, survey weighting and analysis, and data management. She is an expert statistical programmer in R, SAS, and SUDAAN. She earned her PhD in Statistics from Iowa State University and her BS in Statistics from NC State. After earning the RStudio Tidyverse Trainer certification, she co-taught two courses on tidy survey analysis in R. She is currently a Senior Research Statistician at RTI International.

Rebecca J. Powell, PhD, is the Director of Data Management and Survey Programming at Fors Marsh. Her research interests focus on visual design of questionnaires and contact materials. Dr. Powell is a Certified RStudio Tidyverse Trainer and has taught courses and webinars on data management in R and tidy survey analysis in R at both AAPOR and MAPOR. She has a PhD in Survey Research and Methodology from the University of Nebraska-Lincoln and a BS and MS in Applied Statistics from Rochester Institute of Technology.

Isabella C. Velásquez is a content strategist, data enthusiast, and author. Her experience spans data collection, grantmaking, strategy development, and survey instrument development. She earned her MSc in Analytics and BA in Economics and East Asian Languages and Civilizations from the University of Chicago. She currently works as a Sr. Product Marketing Manager at Posit Software, PBC.

Exploring Complex Survey Data Analysis Using R

A Tidy Introduction with {srvyr} and {survey}

Stephanie A. Zimmer, Rebecca J. Powell
and Isabella C. Velásquez

CRC Press
Taylor & Francis Group
Boca Raton London New York

CRC Press is an imprint of the
Taylor & Francis Group, an **informa** business

A CHAPMAN & HALL BOOK

Designed cover image: © Allison Horst

First edition published 2025
by CRC Press
2385 NW Executive Center Drive, Suite 320, Boca Raton FL 33431

and by CRC Press
4 Park Square, Milton Park, Abingdon, Oxon, OX14 4RN

CRC Press is an imprint of Taylor & Francis Group, LLC

Library of Congress Cataloging-in-Publication Data
Names: Zimmer, Stephanie, author. | Powell, Rebecca J., author. |
Velásquez, Isabella, author.
Title: Exploring complex survey data analysis using R : a tidy introduction
with {srvyr} and {survey} / Stephanie Zimmer, Rebecca J. Powell and
Isabella Velásquez.
Description: First edition. | Boca Raton FL : CRC Press, 2025. | Includes
bibliographical references and index.
Identifiers: LCCN 2024022533 (print) | LCCN 2024022534 (ebook) | ISBN
9781032302867 (hardback) | ISBN 9781032306353 (paperback) | ISBN
9781003305996 (ebook)
Subjects: LCSH: R (Computer program language) | Mathematical
statistics--Data processing. | Social surveys--Data processing. |
Sociology--Statistical methods.
Classification: LCC QA276.45.R3 Z56 2025 (print) | LCC QA276.45.R3
(ebook) | DDC 519.50285/5133--dc23/eng20240924
LC record available at https://lccn.loc.gov/2024022533
LC ebook record available at https://lccn.loc.gov/2024022534

ISBN: 978-1-032-30286-7 (hbk)
ISBN: 978-1-032-30635-3 (pbk)
ISBN: 978-1-003-30599-6 (ebk)

DOI: 10.1201/9781003305996

Typeset in Latin Modern
by KnowledgeWorks Global Ltd.

Publisher's note: This book has been prepared from camera-ready copy provided by the authors.

To Will, Tom, and Drew, thanks for all the help with additional chores and plenty of Git consulting!

Contents

IV   Real-life data   185

V   Vignettes   251

List of Figures

List of Tables

Part I

Introduction

1

Introduction

Surveys are valuable tools for gathering information about a population. Researchers, governments, and businesses use surveys to better understand public opinion and behaviors. For example, a non-profit group may analyze societal trends to measure their impact, government agencies may study behaviors to inform policy, or companies may seek to learn customer product preferences to refine business strategy. With survey data, we can explore the world around us.

Surveys are often conducted with a sample of the population. Therefore, to use the survey data to understand the population, we use weights to adjust the survey results for unequal probabilities of selection, nonresponse, and post-stratification. These adjustments ensure the sample accurately represents the population of interest (Gard et al., 2023). To account for the intricate nature of the survey design, analysts rely on statistical software such as SAS, Stata, SUDAAN, and R.

In this book, we focus on R to introduce survey analysis. Our goal is to provide a comprehensive guide for individuals new to survey analysis but with some familiarity with statistics and R programming. We use a combination of the {survey} and {srvyr} packages and present the code following best practices from the tidyverse (Freedman Ellis and Schneider, 2024; Lumley, 2010; Wickham et al., 2019).

1.1 Survey analysis in R

The {survey} package was released on the Comprehensive R Archive Network (CRAN)[1] in 2003 and has been continuously developed over time. This package, primarily authored by Thomas Lumley, offers an extensive array of features, including:

- Calculation of point estimates and estimates of their uncertainty, including means, totals, ratios, quantiles, and proportions

---

[1] https://cran.r-project.org/src/contrib/Archive/survey/

- Estimation of regression models, including generalized linear models, log-linear models, and survival curves
- Variances by Taylor linearization or by replicate weights, including balance repeated replication, jackknife, bootstrap, multistage bootstrap, or user-supplied methods
- Hypothesis testing for means, proportions, and other parameters

The {srvyr} package builds on the {survey} package by providing wrappers for functions that align with the tidyverse philosophy. This is our motivation for using and recommending the {srvyr} package. We find that it is user-friendly for those familiar with the tidyverse packages in R.

For example, while many functions in the {survey} package access variables through formulas, the {srvyr} package uses tidy selection to pass variable names, a common feature in the tidyverse (Henry and Wickham, 2024). Users of the tidyverse are also likely familiar with the magrittr pipe operator (`%>%`), which seamlessly works with functions from the {srvyr} package. Moreover, several common functions from {dplyr}, such as `filter()`, `mutate()`, and `summarize()`, can be applied to survey objects (Wickham et al., 2023a). This enables users to streamline their analysis workflow and leverage the benefits of both the {srvyr} and {tidyverse} packages.

While the {srvyr} package offers many advantages, there is one notable limitation: it doesn't fully incorporate the modeling capabilities of the {survey} package into tidy wrappers. When discussing modeling and hypothesis testing, we primarily rely on the {survey} package. However, we provide information on how to apply the pipe operator to these functions to maintain clarity and consistency in analyses.

1.2   What to expect

This book covers many aspects of survey design and analysis, from understanding how to create design objects to conducting descriptive analysis, statistical tests, and models. We emphasize coding best practices and effective presentation techniques while using real-world data and practical examples to help readers gain proficiency in survey analysis.

Below is a summary of each chapter:

- **Chapter 2 - Overview of surveys**:
    - Overview of survey design processes
    - References for more in-depth knowledge

- **Chapter 3 - Survey data documentation**:
  - Guide to survey documentation types
  - How to read survey documentation
- **Chapter 4 - Getting started**:
  - Installation of packages
  - Introduction to the {srvyrexploR} package and its analytic datasets
  - Outline of the survey analysis process
  - Comparison between the {dplyr} and {srvyr} packages
- **Chapter 5 - Descriptive analyses**:
  - Calculation of point estimates
  - Estimation of standard errors and confidence intervals
  - Calculation of design effects
- **Chapter 6 - Statistical testing**:
  - Statistical testing methods
  - Comparison of means and proportions
  - Goodness-of-fit tests, tests of independence, and tests of homogeneity
- **Chapter 7 - Modeling**:
  - Overview of model formula specifications
  - Linear regression, ANOVA, and logistic regression modeling
- **Chapter 8 - Communication of results**:
  - Strategies for communicating survey results
  - Tools and guidance for creating publishable tables and graphs
- **Chapter 9 - Reproducible research**:
  - Tools and methods for achieving reproducibility
  - Resources for reproducible research
- **Chapter 10 - Sample designs and replicate weights**:
  - Overview of common sampling designs
  - Replicate weight methods
  - How to specify survey designs in R
- **Chapter 11 - Missing data**:
  - Overview of missing data in surveys
  - Approaches to dealing with missing data
- **Chapter 12 - Successful survey analysis recommendations**:
  - Tips for successful analysis
  - Recommendations for debugging
- **Chapter 13 - National Crime Victimization Survey Vignette**:
  - Vignette on analyzing National Crime Victimization Survey (NCVS) data
  - Illustration of analysis requiring multiple files for victimization rates
- **Chapter 14 - AmericasBarometer Vignette**:
  - Vignette on analyzing AmericasBarometer survey data
  - Creation of choropleth maps with survey estimates

The majority of chapters contain code that readers can follow. Each of these chapters starts with a "Prerequisites" section, which includes the code needed to load the packages and datasets used in the chapter. We then provide the main idea of the chapter and examples of how to use the functions. Most chapters conclude with exercises to work through. We provide the solutions to the exercises in the online version of the book[2].

While we provide a brief overview of survey methodology and statistical theory, this book is not intended to be the sole resource for these topics. We reference other materials and encourage readers to seek them out for more information.

1.3   Prerequisites

To get the most out of this book, we assume a survey has already been conducted and readers have obtained a microdata file. Microdata, also known as respondent-level or row-level data, differ from summarized data typically found in tables. Microdata contain individual survey responses, along with analysis weights and design variables such as strata or clusters.

Additionally, the survey data should already include weights and design variables. These are required to accurately calculate unbiased estimates. The concepts and techniques discussed in this book help readers to extract meaningful insights from survey data, but this book does not cover how to create weights, as this is a separate complex topic. If weights are not already created for the survey data, we recommend reviewing other resources focused on weight creation such as Valliant and Dever (2018).

This book is tailored for analysts already familiar with R and the tidyverse, but who may be new to complex survey analysis in R. We anticipate that readers of this book can:

- Install R and their Integrated Development Environment (IDE) of choice, such as RStudio
- Install and load packages from CRAN and GitHub repositories
- Run R code
- Read data from a folder or their working directory
- Understand fundamental tidyverse concepts such as tidy/long/wide data, tibbles, the magrittr pipe (%>%), and tidy selection
- Use the tidyverse packages to wrangle, tidy, and visualize data

If these concepts or skills are unfamiliar, we recommend starting with introductory resources to cover these topics before reading this book. R for Data Science (Wickham et al., 2023c) is a beginner-friendly guide for getting started in data science using R. It offers guidance on preliminary installation steps, basic R syntax, and tidyverse workflows and packages.

---

[2]https://tidy-survey-r.github.io/tidy-survey-book/

1.4 Datasets used in this book

We work with two key datasets throughout the book: the Residential Energy Consumption Survey (RECS – U.S. Energy Information Administration, 2023b) and the American National Election Studies (ANES – DeBell, 2010). We introduce the loading and preparation of these datasets in Chapter 4.

1.5 Conventions

Throughout the book, we use the following typographical conventions:

- Package names are surrounded by curly brackets: {srvyr}
- Function names are in constant-width text format and include parentheses: `survey_mean()`
- Object and variable names are in constant-width text format: `anes_des`

1.6 Getting help

We recommend first trying to resolve errors and issues independently using the tips provided in Chapter 12.

There are several community forums for asking questions, including:

- Posit Community[3]
- R for Data Science Slack Community[4]
- Stack Overflow[5]

Please report any bugs and issues to the book's GitHub repository[6].

---

[3] https://forum.posit.co/
[4] https://rfordatasci.com/
[5] https://stackoverflow.com/
[6] https://github.com/tidy-survey-r/tidy-survey-book/issues

1.7 Acknowledgments

We would like to thank Holly Cast, Greg Freedman Ellis, Joe Murphy, and Sheila Saia for their reviews of the initial draft. Their detailed and honest feedback helped improve this book, and we are grateful for their input. Additionally, this book started with two short courses. The first was at the Annual Conference for the American Association for Public Opinion Research (AAPOR) and the second was a series of webinars for the Midwest Association of Public Opinion Research (MAPOR). We would like to also thank those who assisted us by moderating breakout rooms and answering questions from attendees: Greg Freedman Ellis, Raphael Nishimura, and Benjamin Schneider.

1.8 Colophon

This book was written in bookdown[7] using RStudio[8]. The complete source is available on GitHub[9].

This version of the book was built with R version 4.4.0 (2024-04-24 ucrt) and with the packages listed in Table 1.1.

TABLE 1.1 Package versions and sources used in building this book

Package	Version	Source
DiagrammeR	1.0.11	CRAN
Matrix	1.7–0	CRAN
bookdown	0.39	CRAN
broom	1.0.5	CRAN
censusapi	0.9.0.9000	GitHub (hrecht/censusapi@74334d4)
dplyr	1.1.4	CRAN
forcats	1.0.0	CRAN
ggpattern	1.0.1	CRAN
ggplot2	3.5.1	CRAN
gt	0.11.0.9000	GitHub (rstudio/gt@28de628)
gtsummary	1.7.2	CRAN
haven	2.5.4	CRAN
janitor	2.2.0	CRAN
kableExtra	1.4.0	CRAN

[7]http://bookdown.org/
[8]http://www.rstudio.com/ide/
[9]https://github.com/tidy-survey-r/tidy-survey-book

knitr	1.46	CRAN
labelled	2.13.0	CRAN
lubridate	1.9.3	CRAN
naniar	1.1.0	CRAN
osfr	0.2.9	CRAN
prettyunits	1.2.0	CRAN
purrr	1.0.2	CRAN
readr	2.1.5	CRAN
renv	1.0.7	CRAN
rmarkdown	2.26	CRAN
rnaturalearth	1.0.1	CRAN
rnaturalearthdata	1.0.0	CRAN
sf	1.0–16	CRAN
srvyr	1.3.0	CRAN
srvyrexploR	1.0.1	GitHub (tidy-survey-r/srvyrexploR@cdf9316)
stringr	1.5.1	CRAN
styler	1.10.3	CRAN
survey	4.4–2	CRAN
survival	3.6–4	CRAN
tibble	3.2.1	CRAN
tidycensus	1.6.3	CRAN
tidyr	1.3.1	CRAN
tidyselect	1.2.1	CRAN
tidyverse	2.0.0	CRAN

2

Overview of surveys

2.1 Introduction

Developing surveys to gather accurate information about populations involves an intricate and time-intensive process. Researchers can spend months, or even years, developing the study design, questions, and other methods for a single survey to ensure high-quality data is collected.

Before analyzing survey data, we recommend understanding the entire survey life cycle. This understanding can provide better insight into what types of analyses should be conducted on the data. The survey life cycle consists of the necessary stages to execute a survey project successfully. Each stage influences the survey's timing, costs, and feasibility, consequently impacting the data collected and how we should analyze them. Figure 2.1 shows a high-level overview of the survey process.

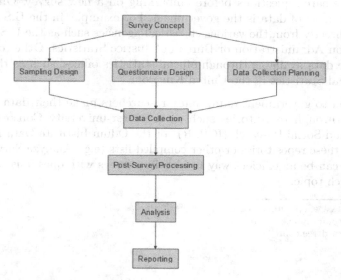

FIGURE 2.1 Overview of the survey process

The survey life cycle starts with a research topic or question of interest (e.g., the impact that childhood trauma has on health outcomes later in life). Drawing from available resources can result in a reduced burden on respondents, lower costs, and faster research outcomes. Therefore, we recommend reviewing existing data sources to determine if data that can address this question are already available. However, if existing data cannot answer the nuances of the research question, we can capture the exact data we need through a questionnaire, or a set of questions.

To gain a deeper understanding of survey design and implementation, we recommend reviewing several pieces of existing literature in detail (e.g., Biemer and Lyberg, 2003; Bradburn et al., 2004; Dillman et al., 2014; Groves et al., 2009; Tourangeau et al., 2000; Valliant et al., 2013).

2.2 Searching for public-use survey data

Throughout this book, we use public-use datasets from different surveys, including the American National Election Studies (ANES), the Residential Energy Consumption Survey (RECS), the National Crime Victimization Survey (NCVS), and the AmericasBarometer surveys.

As mentioned above, we should look for existing data that can provide insights into our research questions before embarking on a new survey. One of the greatest sources of data is the government. For example, in the U.S., we can get data directly from the various statistical agencies such as the U.S. Energy Information Administration or Bureau of Justice Statistics. Other countries often have data available through official statistics offices, such as the Office for National Statistics in the United Kingdom.

In addition to government data, many researchers make their data publicly available through repositories such as the Inter-university Consortium for Political and Social Research (ICPSR)[1] or the Odum Institute Data Archive[2]. Searching these repositories or other compiled lists (e.g., Analyze Survey Data for Free[3]) can be an efficient way to identify surveys with questions related to our research topic.

---

[1] https://www.icpsr.umich.edu/web/pages/ICPSR/ssvd/
[2] https://odum.unc.edu/archive/
[3] https://asdfree.com

2.3 Pre-survey planning

There are multiple things to consider when starting a survey. Errors are the differences between the true values of the variables being studied and the values obtained through the survey. Each step and decision made before the launch of the survey impact the types of errors that are introduced into the data, which in turn impact how to interpret the results.

Generally, survey researchers consider there to be seven main sources of error that fall under either Representation or Measurement (Groves et al., 2009):

- Representation
  - Coverage Error: A mismatch between the population of interest and the sampling frame, the list from which the sample is drawn.
  - Sampling Error: Error produced when selecting a sample, the subset of the population, from the sampling frame. This error is due to randomization, and we discuss how to quantify this error in Chapter 10. There is no sampling error in a census, as there is no randomization. The sampling error measures the difference between all potential samples under the same sampling method.
  - Nonresponse Error: Differences between those who responded and did not respond to the survey (unit nonresponse) or a given question (item nonresponse).
  - Adjustment Error: Error introduced during post-survey statistical adjustments.
- Measurement
  - Validity: A mismatch between the research topic and the question(s) used to collect that information.
  - Measurement Error: A mismatch between what the researcher asked and how the respondent answered.
  - Processing Error: Edits by the researcher to responses provided by the respondent (e.g., adjustments to data based on illogical responses).

Almost every survey has errors. Researchers attempt to conduct a survey that reduces the total survey error, or the accumulation of all errors that may arise throughout the survey life cycle. By assessing these different types of errors together, researchers can seek strategies to maximize the overall survey quality and improve the reliability and validity of results (Biemer, 2010). However, attempts to reduce individual source errors (and therefore total survey error) come at the price of time and money. For example:

- Coverage Error Tradeoff: Researchers can search for or create more accurate and updated sampling frames, but they can be difficult to construct or obtain.

- Sampling Error Tradeoff: Researchers can increase the sample size to reduce sampling error; however, larger samples can be expensive and time-consuming to field.
- Nonresponse Error Tradeoff: Researchers can increase or diversify efforts to improve survey participation, but this may be resource-intensive while not entirely removing nonresponse bias.
- Adjustment Error Tradeoff: Weighting is a statistical technique used to adjust the contribution of individual survey responses to the final survey estimates. It is typically done to make the sample more representative of the population of interest. However, if researchers do not carefully execute the adjustments or base them on inaccurate information, they can introduce new biases, leading to less accurate estimates.
- Validity Error Tradeoff: Researchers can increase validity through a variety of ways, such as using established scales or collaborating with a psychometrician during survey design to pilot and evaluate questions. However, doing so increases the amount of time and resources needed to complete survey design.
- Measurement Error Tradeoff: Researchers can use techniques such as questionnaire testing and cognitive interviewing to ensure respondents are answering questions as expected. However, these activities require time and resources to complete.
- Processing Error Tradeoff: Researchers can impose rigorous data cleaning and validation processes. However, this requires supervision, training, and time.

The challenge for survey researchers is to find the optimal tradeoffs among these errors. They must carefully consider ways to reduce each error source and total survey error while balancing their study's objectives and resources.

For survey analysts, understanding the decisions that researchers took to minimize these error sources can impact how results are interpreted. The remainder of this chapter explores critical considerations for survey development. We explore how to consider each of these sources of error and how these error sources can inform the interpretations of the data.

2.4   Study design

From formulating methodologies to choosing an appropriate sampling frame, the study design phase is where the blueprint for a successful survey takes shape. Study design encompasses multiple parts of the survey life cycle, including decisions on the population of interest, survey mode (the format through which a survey is administered to respondents), timeline, and questionnaire design. Knowing who and how to survey individuals depends on the study's goals and the feasibility of implementation. This section explores the strategic planning that lays the foundation for a survey.

2.4.1 Sampling design

The set or group we want to survey is known as the population of interest or the target population. The population of interest could be broad, such as "all adults age 18+ living in the U.S." or a specific population based on a particular characteristic or location. For example, we may want to know about "adults aged 18–24 who live in North Carolina" or "eligible voters living in Illinois."

However, a sampling frame with contact information is needed to survey individuals in these populations of interest. If we are looking at eligible voters, the sampling frame could be the voting registry for a given state or area. If we are looking at more board populations of interest, like all adults in the United States, the sampling frame is likely imperfect. In these cases, a full list of individuals in the United States is not available for a sampling frame. Instead, we may choose to use a sampling frame of mailing addresses and send the survey to households, or we may choose to use random digit dialing (RDD) and call random phone numbers (that may or may not be assigned, connected, and working).

These imperfect sampling frames can result in coverage error where there is a mismatch between the population of interest and the list of individuals we can select. For example, if we are looking to obtain estimates for "all adults aged 18+ living in the U.S.," a sampling frame of mailing addresses will miss specific types of individuals, such as the homeless, transient populations, and incarcerated individuals. Additionally, many households have more than one adult resident, so we would need to consider how to get a specific individual to fill out the survey (called within household selection) or adjust the population of interest to report on "U.S. households" instead of "individuals."

Once we have selected the sampling frame, the next step is determining how to select individuals for the survey. In rare cases, we may conduct a census and survey everyone on the sampling frame. However, the ability to implement a questionnaire at that scale is something only a few can do (e.g., government censuses). Instead, we typically choose to sample individuals and use weights to estimate numbers in the population of interest. They can use a variety of different sampling methods, and more information on these can be found in Chapter 10. This decision of which sampling method to use impacts sampling error and can be accounted for in weighting.

Example: Number of pets in a household

Let's use a simple example where we are interested in the average number of pets in a household. We need to consider the population of interest for this study. Specifically, are we interested in all households in a given country or households in a more local area (e.g., city or state)? Let's assume we are interested in the number of pets in a U.S. household with at least one adult (18 years or older). In this case, a sampling frame of mailing addresses would introduce only a small amount of coverage error as the frame would closely match our population of interest. Specifically, we would likely want to use the Computerized Delivery Sequence File (CDSF), which is a file of mailing addresses that the United States Postal Service (USPS) creates and covers nearly 100% of U.S. households (Harter et al., 2016). To sample these households, for simplicity, we use a stratified simple random sample design (see Chapter 10 for more information on sample designs), where we randomly sample households within each state (i.e., we stratify by state).

Throughout this chapter, we build on this example research question to plan a survey.

2.4.2 Data collection planning

With the sampling design decided, researchers can then decide how to survey these individuals. Specifically, the modes used for contacting and surveying the sample, how frequently to send reminders and follow-ups, and the overall timeline of the study are some of the major data collection determinations. Traditionally, survey researchers have considered there to be four main modes[4]:

- Computer-Assisted Personal Interview (CAPI; also known as face-to-face or in-person interviewing)
- Computer-Assisted Telephone Interview (CATI; also known as phone or telephone interviewing)
- Computer-Assisted Web Interview (CAWI; also known as web or online interviewing)
- Paper and Pencil Interview (PAPI)

We can use a single mode to collect data or multiple modes (also called mixed-modes). Using mixed-modes can allow for broader reach and increase response rates depending on the population of interest (Biemer et al., 2017; DeLeeuw, 2005, 2018). For example, we could both call households to conduct a CATI survey and send mail with a PAPI survey to the household. By using both modes, we could gain participation through the mail from individuals who do not pick up the phone to unknown numbers or through the phone from individuals who do not open all of their mail. However, mode effects (where responses differ based on the mode of response) can be present in the data and may need to be considered during analysis.

When selecting which mode, or modes, to use, understanding the unique aspects of the chosen population of interest and sampling frame provides insight into

---

[4]Other modes such as using mobile apps or text messaging can also be considered, but at the time of publication, they have smaller reach or are better for longitudinal studies (i.e., surveying the same individuals over many time periods of a single study).

how they can best be reached and engaged. For example, if we plan to survey adults aged 18–24 who live in North Carolina, asking them to complete a survey using CATI (i.e., over the phone) would likely not be as successful as other modes like the web. This age group does not talk on the phone as much as other generations and often does not answer phone calls from unknown numbers. Additionally, the mode for contacting respondents relies on what information is available in the sampling frame. For example, if our sampling frame includes an email address, we could email our selected sample members to convince them to complete a survey. Alternatively, if the sampling frame is a list of mailing addresses, we could contact sample members with a letter.

It is important to note that there can be a difference between the contact and survey modes. For example, if we have a sampling frame with addresses, we can send a letter to our sample members and provide information on completing a web survey. Another option is using mixed-mode surveys by mailing sample members a paper and pencil survey but also including instructions to complete the survey online. Combining different contact modes and different survey modes can be helpful in reducing unit nonresponse error–where the entire unit (e.g., a household) does not respond to the survey at all–as different sample members may respond better to different contact and survey modes. However, when considering which modes to use, it is important to make access to the survey as easy as possible for sample members to reduce burden and unit nonresponse.

Another way to reduce unit nonresponse error is by varying the language of the contact materials (Dillman et al., 2014). People are motivated by different things, so constantly repeating the same message may not be helpful. Instead, mixing up the messaging and the type of contact material the sample member receives can increase response rates and reduce the unit nonresponse error. For example, instead of only sending standard letters, we could consider sending mailings that invoke "urgent" or "important" thoughts by sending priority letters or using other delivery services like FedEx, UPS, or DHL.

A study timeline may also determine the number and types of contacts. If the timeline is long, there is plentiful time for follow-ups and diversified messages in contact materials. If the timeline is short, then fewer follow-ups can be implemented. Many studies start with the tailored design method put forth by Dillman et al. (2014) and implement five contacts:

- Pre-notification (Pre-notice) to let sample members know the survey is coming
- Invitation to complete the survey
- Reminder to also thank the respondents who have already completed the survey
- Reminder (with a replacement paper survey if needed)
- Final reminder

This method is easily adaptable based on the study timeline and needs but provides a starting point for most studies.

Example: Number of pets in a household

Let's return to our example of the average number of pets in a household. We are using a sampling frame of mailing addresses, so we recommend starting our data collection with letters mailed to households, but later in data collection, we want to send interviewers to the house to conduct an in-person (or CAPI) interview to decrease unit nonresponse error. This means we have two contact modes (paper and in-person). As mentioned above, the survey mode does not have to be the same as the contact mode, so we recommend a mixed-mode study with both web and CAPI modes. Let's assume we have 6 months for data collection, so we could recommend Table 2.1's protocol:

TABLE 2.1 Protocol example for 6-month web and CAPI data collection

Week	Contact Mode	Contact Message	Survey Mode Offered
1	Mail: Letter	Pre-notice	—
2	Mail: Letter	Invitation	Web
3	Mail: Postcard	Thank You/Reminder	Web
6	Mail: Letter in large envelope	Animal Welfare Discussion	Web
10	Mail: Postcard	Inform Upcoming In-Person Visit	Web
14	In-Person Visit	—	CAPI
16	Mail: Letter	Reminder of In-Person Visit	Web, but includes a number to call to schedule CAPI
20	In-Person Visit	—	CAPI
25	Mail: Letter in large envelope	Survey Closing Notice	Web, but includes a number to call to schedule CAPI

This is just one possible protocol that we can use that starts respondents with the web (typically done to reduce costs). However, we could begin in-person data collection earlier during the data collection period or ask interviewers to attempt more than two visits with a household.

2.4.3 Questionnaire design

When developing the questionnaire, it can be helpful to first outline the topics to be asked and include the "why" each question or topic is important to the research question(s). This can help us better tailor the questionnaire and reduce the number of questions (and thus the burden on the respondent) if topics are deemed irrelevant to the research question. When making these decisions, we should also consider questions needed for weighting. While we would love to have everyone in our population of interest answer our survey, this rarely happens. Thus, including questions about demographics in the survey can assist with weighting for nonresponse errors (both unit and item nonresponse). Knowing the details of the sampling plan and what may impact coverage error and sampling error can help us determine what types of demographics to include. Thus questionnaire design is typically done in conjunction with sampling design.

We can benefit from the work of others by using questions from other surveys. Demographic sections in surveys, such as race, ethnicity, or education, often are borrowed questions from a government census or other official surveys. Question banks such as the ICPSR variable search[5] can provide additional potential questions.

If a question does not exist in a question bank, we can craft our own. When developing survey questions, we should start with the research topic and attempt to write questions that match the concept. The closer the question asked is to the overall concept, the better validity there is. For example, if we want to know how people consume T.V. series and movies but only ask a question about how many T.V.s are in the house, then we would be missing other ways that people watch T.V. series and movies, such as on other devices or at places outside of the home. As mentioned above, we can employ techniques to increase the validity of questionnaires. For example, questionnaire testing involves piloting the survey instrument to identify and fix potential issues before conducting the main survey. Additionally, we could conduct cognitive interviews – a technique where we walk through the survey with participants, encouraging them to speak their thoughts out loud to uncover how they interpret and understand survey questions.

Additionally, when designing questions, we should consider the mode for the survey and adjust the language appropriately. In self-administered surveys (e.g., web or mail), respondents can see all the questions and response options, but that is not the case in interviewer-administered surveys (e.g., CATI or CAPI). With interviewer-administered surveys, the response options must be read aloud to the respondents, so the question may need to be adjusted to create a better flow to the interview. Additionally, with self-administered surveys, because the respondents are viewing the questionnaire, the formatting of the questions is even more critical to ensure accurate measurement. Incorrect formatting or wording can result in measurement error, so following best practices or using existing validated questions can reduce error. There are multiple resources to help researchers draft questions for different modes (e.g., Bradburn et al., 2004; Dillman et al., 2014; Fowler and Mangione, 1989; Tourangeau et al., 2004).

---

[5]https://www.icpsr.umich.edu/web/pages/ICPSR/ssvd/

Example: Number of pets in a household

As part of our survey on the average number of pets in a household, we may want to know what animal most people prefer to have as a pet. Let's say we have a question in our survey as displayed in Figure 2.2.

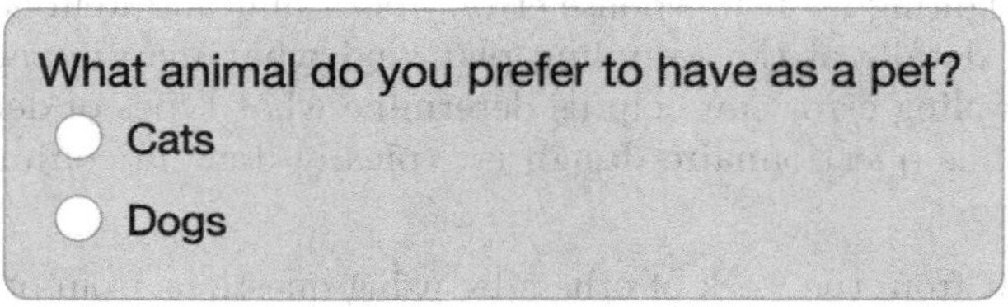

> **What animal do you prefer to have as a pet?**
> ○ Cats
> ○ Dogs

FIGURE 2.2 Example question asking pet preference type

This question may have validity issues as it only provides the options of "dogs" and "cats" to respondents, and the interpretation of the data could be incorrect. For example, if we had 100 respondents who answered the question and 50 selected dogs, then the results of this question cannot be "50% of the population prefers to have a dog as a pet," as only two response options were provided. If a respondent taking our survey prefers turtles, they could either be forced to choose a response between these two (i.e., interpret the question as "between dogs and cats, which do you prefer?" and result in measurement error), or they may not answer the question (which results in item nonresponse error). Based on this, the interpretation of this question should be, "When given a choice between dogs and cats, 50% of respondents preferred to have a dog as a pet."

To avoid this issue, we should consider these possibilities and adjust the question accordingly. One simple way could be to add an "other" response option to give respondents a chance to provide a different response. The "other" response option could then include a way for respondents to write their other preference. For example, we could rewrite this question as displayed in Figure 2.3.

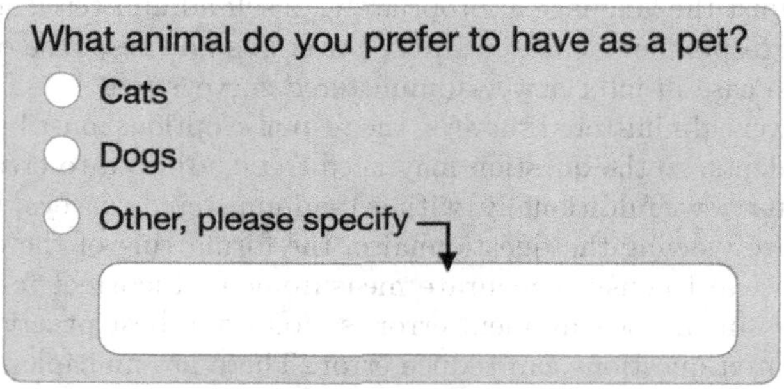

> **What animal do you prefer to have as a pet?**
> ○ Cats
> ○ Dogs
> ○ Other, please specify ⌐
> [   ]

FIGURE 2.3 Example question asking pet preference type with other specify option

We can then code the responses from the open-ended box and get a better understanding of the respondent's choice of preferred pet. Interpreting this question becomes easier as researchers no longer need to qualify the results with the choices provided.

This is a simple example of how the presentation of the question and options can impact the findings. For more complex topics and questions, we must thoroughly consider how to mitigate any impacts from the presentation, formatting, wording, and other aspects. For survey analysts, reviewing not only the data but also the wording of the questions is crucial to ensure the results are presented in a manner consistent with the question asked. Chapter 3 provides further details on how to review existing survey documentation to inform our analyses, and Chapter 8 goes into more details on communicating results.

2.5 Data collection

Once the data collection starts, we try to stick to the data collection protocol designed during pre-survey planning. However, effective researchers also prepare to adjust their plans and adapt as needed to the current progress of data collection (Schouten et al., 2018). Some extreme examples could be natural disasters that could prevent mailings or interviewers from getting to the sample members. This could cause an in-person survey needing to quickly pivot to a self-administered survey, or the field period could be delayed, for example. Others could be smaller in that something newsworthy occurs connected to the survey, so we could choose to play this up in communication materials. In addition to these external factors, there could be factors unique to the survey, such as lower response rates for a specific subgroup, so the data collection protocol may need to find ways to improve response rates for that specific group.

2.6 Post-survey processing

After data collection, various activities need to be completed before we can analyze the survey. Multiple decisions made during this post-survey phase can assist us in reducing different error sources, such as weighting to account for the sample selection. Knowing the decisions made in creating the final analytic data can impact how we use the data and interpret the results.

2.6.1   Data cleaning and imputation

Post-survey cleaning is one of the first steps we do to get the survey responses
into an analytic dataset. Data cleaning can consist of correcting inconsistent
data (e.g., with skip pattern errors or multiple questions throughout the survey
being consistent with each other), editing numeric entries or open-ended
responses for grammar and consistency, or recoding open-ended questions into
categories for analysis. There is no universal set of fixed rules that every survey
must adhere to. Instead, each survey or research study should establish its
own guidelines and procedures for handling various cleaning scenarios based
on its specific objectives.

We should use our best judgment to ensure data integrity, and all decisions
should be documented and available to those using the data in the analysis.
Each decision we make impacts processing error, so often, multiple people
review these rules or recode open-ended data and adjudicate any differences in
an attempt to reduce this error.

Another crucial step in post-survey processing is imputation. Often, there
is item nonresponse where respondents do not answer specific questions. If
the questions are crucial to analysis efforts or the research question, we may
implement imputation to reduce item nonresponse error. Imputation is a
technique for replacing missing or incomplete data values with estimated
values. However, as imputation is a way of assigning values to missing data
based on an algorithm or model, it can also introduce processing error, so we
should consider the overall implications of imputing data compared to having
item nonresponse. There are multiple ways to impute data. We recommend
reviewing other resources like Kim and Shao (2021) for more information.

Example: Number of pets in a household

Let's return to the question we created to ask about animal preference. The
"other specify" invites respondents to specify the type of animal they prefer
to have as a pet. If respondents entered answers such as "puppy," "turtle,"
"rabit," "rabbit," "bunny," "ant farm," "snake," "Mr. Purr," then we may wish
to categorize these write-in responses to help with analysis. In this example,
"puppy" could be assumed to be a reference to a "Dog" and could be recoded
there. The misspelling of "rabit" could be coded along with "rabbit" and
"bunny" into a single category of "Bunny or Rabbit." These are relatively
standard decisions that we can make. The remaining write-in responses could
be categorized in a few different ways. "Mr. Purr," which may be someone's
reference to their own cat, could be recoded as "Cat," or it could remain as
"Other" or some category that is "Unknown." Depending on the number of
responses related to each of the others, they could all be combined into a single
"Other" category, or maybe categories such as "Reptiles" or "Insects" could
be created. Each of these decisions may impact the interpretation of the data,

so we should document the types of responses that fall into each of the new categories and any decisions made.

2.6.2 Weighting

We can address some error sources identified in the previous sections using weighting. During the weighting process, weights are created for each respondent record. These weights allow the survey responses to generalize to the population. A weight, generally, reflects how many units in the population each respondent represents. Often, the weight is constructed such that the sum of the weights is the size of the population.

Weights can address coverage, sampling, and nonresponse errors. Many published surveys include an "analysis weight" variable that combines these adjustments. However, weighting itself can also introduce adjustment error, so we need to balance which types of errors should be corrected with weighting. The construction of weights is outside the scope of this book, we recommend referencing other materials if interested in weight construction (Valliant and Dever, 2018). Instead, this book assumes the survey has been completed, weights are constructed, and data are available to users.

Example: Number of pets in a household

In the simple example of our survey, we decided to obtain a random sample from each state to select our sample members. Knowing this sampling design, we can include selection weights for analysis that account for how the sample members were selected for the survey. Additionally, the sampling frame may have the type of building associated with each address, so we could include the building type as a potential nonresponse weighting variable, along with some interviewer observations that may be related to our research topic of the average number of pets in a household. Combining these weights, we can create an analytic weight that analysts need to use when analyzing the data.

2.6.3 Disclosure

Before data is released publicly, we need to ensure that individual respondents cannot be identified by the data when confidentiality is required. There are a variety of different methods that can be used. Here we describe a few of the most commonly used:

- Data swapping: We may swap specific data values across different respondents so that it does not impact insights from the data but ensures that specific individuals cannot be identified.
- Top/bottom coding: We may choose top or bottom coding to mask extreme values. For example, we may top-code income values such that households with income greater than $500,000 are coded as "$500,000 or more" with

other incomes being presented as integers between \$0 and \$499,999. This can impact analyses at the tails of the distribution.

- Coarsening: We may use coarsening to mask unique values. For example, a survey question may ask for a precise income but the public data may include income as a categorical variable. Another example commonly used in survey practice is to coarsen geographic variables. Data collectors likely know the precise address of sample members, but the public data may only include the state or even region of respondents.
- Perturbation: We may add random noise to outcomes. As with swapping, this is done so that it does not impact insights from the data but ensures that specific individuals cannot be identified.

There is as much art as there is science to the methods used for disclosure. Only high-level comments about the disclosure are provided in the survey documentation, not specific details. This ensures nobody can reverse the disclosure and thus identify individuals. For more information on different disclosure methods, please see Skinner (2009) and the AAPOR Standards[6].

2.6.4 Documentation

Documentation is a critical step of the survey life cycle. We should systematically record all the details, decisions, procedures, and methodologies to ensure transparency, reproducibility, and the overall quality of survey research.

Proper documentation allows analysts to understand, reproduce, and evaluate the study's methods and findings. Chapter 3 dives into how analysts should use survey data documentation.

2.7 Post-survey data analysis and reporting

After completing the survey life cycle, the data are ready for analysts. Chapter 4 continues from this point. For more information on the survey life cycle, please explore the references cited throughout this chapter.

---

[6]https://aapor.org/standards-and-ethics/disclosure-standards/

3

Survey data documentation

3.1 Introduction

Survey documentation helps us prepare before we look at the actual survey data. The documentation includes technical guides, questionnaires, codebooks, errata, and other useful resources. By taking the time to review these materials, we can gain a comprehensive understanding of the survey data (including research and design decisions discussed in Chapters 2 and 10) and conduct our analysis more effectively.

Survey documentation can vary in organization, type, and ease of use. The information may be stored in any format—PDFs, Excel spreadsheets, Word documents, and so on. Some surveys bundle documentation together, such as providing the codebook and questionnaire in a single document. Others keep them in separate files. Despite these variations, we can gain a general understanding of the documentation types and what aspects to focus on in each.

3.2 Types of survey documentation

3.2.1 Technical documentation

The technical documentation, also known as user guides or methodology/analysis guides, highlights the variables necessary to specify the survey design. We recommend concentrating on these key sections:

- Introduction: The introduction orients us to the survey. This section provides the project's background, the study's purpose, and the main research questions.
- Study design: The study design section describes how researchers prepared and administered the survey.
- Sample: The sample section describes the sample frame, any known sampling errors, and limitations of the sample. This section can contain recommendations on how to use sampling weights. Look for weight information,

whether the survey design contains strata, clusters/PSUs, or replicate weights. Also, look for population sizes, finite population correction, or replicate weight scaling information. Additional detail on sample designs is available in Chapter 10.

- Notes on fielding: Any additional notes on fielding, such as response rates, may be found in the technical documentation.

The technical documentation may include other helpful resources. For example, some technical documentation includes syntax for SAS, SUDAAN, Stata, and/or R, so we do not have to create this code from scratch.

3.2.2   Questionnaires

A questionnaire is a series of questions used to collect information from people in a survey. It can ask about opinions, behaviors, demographics, or even just numbers like the count of lightbulbs, square footage, or farm size. Questionnaires can employ different types of questions, such as closed-ended (e.g., select one or check all that apply), open-ended (e.g., numeric or text), Likert scales (e.g., a 5- or 7-point scale specifying a respondent's level of agreement to a statement), or ranking questions (e.g., a list of options that a respondent ranks by preference). It may randomize the display order of responses or include instructions that help respondents understand the questions. A survey may have one questionnaire or multiple, depending on its scale and scope.

The questionnaire is another important resource for understanding and interpreting the survey data (see Section 2.4.3), and we should use it alongside any analysis. It provides details about each of the questions asked in the survey, such as question name, question wording, response options, skip logic, randomizations, display specifications, mode differences, and the universe (the subset of respondents who were asked a question).

In Figure 3.1, we show an example from the American National Election Studies (ANES) 2020 questionnaire (American National Election Studies, 2021). The figure shows the question name (POSTVOTE_RVOTE), description (Did R Vote?), full wording of the question and responses, response order, universe, question logic (this question was only asked if vote_pre = 0), and other specifications. The section also includes the variable name, which we can link to the codebook.

The content and structure of questionnaires vary depending on the specific survey. For instance, question names may be informative (like the ANES example above), sequential, or denoted by a code. In some cases, surveys may not use separate names for questions and variables. Figure 3.2 shows an example from the Behavioral Risk Factor Surveillance System (BRFSS) questionnaire that shows a sequential question number and a coded variable name (as opposed to a question name) (Centers for Disease Control and Prevention (CDC), 2021).

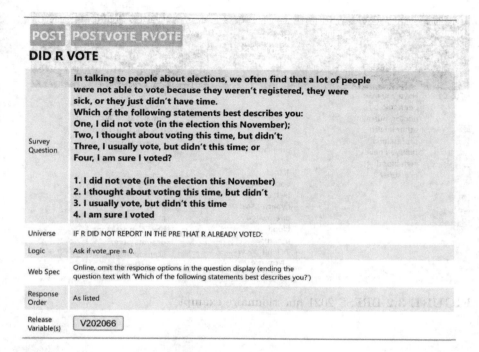

FIGURE 3.1 ANES 2020 questionnaire example

We should factor in the details of a survey when conducting our analyses. For example, surveys that use various modes (e.g., web and mail) may have differences in question wording or skip logic, as web surveys can include fills or automate skip logic. If large enough, these variations could warrant separate analyses for each mode.

3.2.3 Codebooks

While a questionnaire provides information about the questions posed to respondents, the codebook explains how the survey data were coded and recorded. It lists details such as variable names, variable labels, variable meanings, codes for missing data, value labels, and value types (whether categorical, continuous, etc.). The codebook helps us understand and use the variables appropriately in our analysis. In particular, the codebook (as opposed to the questionnaire) often includes information on missing data. Note that the term data dictionary is sometimes used interchangeably with codebook, but a data dictionary may include more details on the structure and elements of the data.

Figure 3.3 is a question from the ANES 2020 codebook (American National Election Studies, 2022). This section indicates a variable's name (V202066),

Question Number	Question text	Variable names	Responses (DO NOT READ UNLESS OTHERWISE NOTED)	SKIP INFO/ CATI Note	Interviewer Note (s)	Column(s)
C05.01	Have you ever been told by a doctor, nurse, or other health professional that you have high blood pressure?	BPHIGH6	1 Yes 2 Yes, but female told only during pregnancy 3 No 4 Told borderline high or pre-hypertensive or elevated blood pressure 7 Don't know / Not sure 9 Refused	Go to next section	If "Yes" and respondent is female, ask: "Was this only when you were pregnant?" By other health professional we mean nurse practitioner, a physician assistant, or some other licensed health professional.	

FIGURE 3.2 BRFSS 2021 questionnaire example

V202066 POST: DID R VOTE IN NOVEMBER 2020 ELECTION

Question	In talking to people about elections, we often find that a lot of peoplewere not able to vote because they weren't registered, they weresick, or they just didn't have time. Which of the following statements best describes you:
Value Labels	-9. Refused -7. No post-election data, deleted due to incomplete interview -6. No post-election interview -1. Inapplicable 1. I did not vote (in the election this November) 2. I thought about voting this time, but didn't 3. I usually vote, but didn't this time 4. I am sure I voted
Universe	IF R DID NOT REPORT IN THE PRE THAT R ALREADY VOTED:
Survey Question(s)	POSTVOTE_RVOTE

FIGURE 3.3 ANES 2020 codebook example

question wording, value labels, universe, and associated survey question (POSTVOTE_RVOTE).

Reviewing the questionnaires and codebooks in parallel can clarify how to interpret the variables (Figures 3.1 and 3.3), as questions and variables do not always correspond directly to each other in a one-to-one mapping. A single question may have multiple associated variables, or a single variable may summarize multiple questions.

3.2.4 Errata

An erratum (singular) or errata (plural) is a document that lists errors found in a publication or dataset. The purpose of an erratum is to correct or update inaccuracies in the original document. Examples of errata include:

- Issuing a corrected data table after realizing a typo or mistake in a table cell
- Reporting incorrectly programmed skips in an electronic survey where questions are skipped by the respondent when they should not have been

For example, the 2004 ANES dataset released an erratum, notifying analysts to remove a specific row from the data file due to the inclusion of a respondent who should not have been part of the sample. Adhering to an issued erratum helps us increase the accuracy and reliability of analysis.

3.2.5 Additional resources

Survey documentation may include additional material, such as interviewer instructions or "show cards" provided to respondents during interviewer-administered surveys to help respondents answer questions. Explore the survey website to find out what resources were used and in what contexts.

3.3 Missing data coding

Some observations in a dataset may have missing data. This can be due to design or nonresponse, and these concepts are detailed in Chapter 11. In that chapter, we also discuss how to analyze data with missing values. This chapter walks through how to understand documentation related to missing data.

The survey documentation, often the codebook, represents the missing data with a code. The codebook may list different codes depending on why certain data points are missing. In the example of variable V202066 from the ANES (Figure 3.3), -9 represents "Refused," -7 means that the response was deleted due to an incomplete interview, -6 means that there is no response because there was no follow-up interview, and `-1` means "Inapplicable" (due to a designed skip pattern).

As another example, there may be a summary variable that describes the missingness of a set of variables — particularly with "select all that apply" or "multiple response" questions. In the National Crime Victimization Survey (NCVS), respondents who are victims of a crime and saw the offender are asked if the offender had a weapon and then asked what the type of weapon was. This part of the questionnaire from 2021 is shown in Figure 3.4 (U. S. Bureau of Justice Statistics, 2020).

For these multiple response variables (select all that apply), the NCVS codebook includes what they call a "lead-in" variable that summarizes the response. This lead-in variable provides metadata information on how a respondent answered the question. For example, question 23a on the weapon type, the lead-in variable is V4050 (shown in Figure 3.5) indicates the quality and type of response (U. S. Bureau of Justice Statistics, 2022). In the codebook, this variable is then followed by a set of variables for each weapon type. An example of one of the individual variables from the codebook, the handgun (V4051), is shown in Figure 3.6 (U. S. Bureau of Justice Statistics, 2022). We will dive into how to analyze this variable in Chapter 11.

When data are read into R, some values may be system missing, that is they are coded as `NA` even if that is not evident in a codebook. We discuss in Chapter 11 how to analyze data with `NA` values and review how R handles missing data in calculations.

22. WEAPON PRESENT	637	1 ☐ Yes - ASK 23a	
Did the offender have a weapon such as a gun or knife, or something to use as a weapon, such as a bottle or wrench?		2 ☐ No .... } SKIP to 24 3 ☐ Don't know	
23a. WEAPON	638	1 ☐ Hand gun (pistol, revolver, etc.)	
What was the weapon?		2 ☐ Other gun (rifle, shotgun, etc.) .... 3 ☐ Knife	SKIP to 24
Probe: **Anything else?** Enter all that apply.		4 ☐ Other sharp object (scissors, ice pick, axe, etc.) .... 5 ☐ Blunt object (rock, club, blackjack, etc.) ...... 6 ☐ Other - Specify - ASK 23b	

FIGURE 3.4 Excerpt from the NCVS 2020-2021 Crime Incident Report - Weapon Type

V4050 - LI WHAT WAS WEAPON

Location: 152-152 (width: 1; decimal: 0) (Incident Record-Type File)

Location: 652-652 (width: 1; decimal: 0) (2021 Collection Year Incident-Level Extract File)

Variable Type: numeric

Question:

What was the weapon? Anything else?
Text:
Source code: 638[L]

Lead-in code

(Summary of single response entries for multiple response question. Detailed responses are given in VARS V4051-V4058)

Value	Label
1	At least one good entry in one or more of the answer category codes 1-7
3	Indicates"Yes-Type Weapon-NA". Zeroes will appear in each of the category codes and an 8 will appear in the residue code for this situation
7	Indicates"Gun Type Unknown". A 1 will always appear in V4057 and an 8 in V4058 if the only weapon reported is"Gun Type Unknown". If other weapons are also reported, a 0 will appear in V4058.
8	No good entry (out-of-range) in any of the answer category codes 1-7 or item was blank
	Missing Data
9	Out of universe

FIGURE 3.5 Excerpt from the NCVS 2021 Codebook for V4050 - LI WHAT WAS WEAPON

V4051 - C WEAPON: HAND GUN

Location: 153-153 (width: 1; decimal: 0) (Incident Record-Type File)

Location: 653-653 (width: 1; decimal: 0) (2021 Collection Year Incident-Level Extract File)

Variable Type: numeric

Question:

Handgun present (pistol, revolver, etc.)

What was the weapon? Anything else?
Text:
Source code: 638[1]

Value	Label
0	No
1	Yes
8	Residue
	Missing Data
9	Out of universe

FIGURE 3.6 Excerpt from the NCVS 2021 Codebook for V4051 - C WEAPON: HAND GUN

3.4   Example: ANES 2020 survey documentation

Let's look at the survey documentation for the ANES 2020 and the documentation from their website[1]. Navigating to "User Guide and Codebook" (American National Election Studies, 2022), we can download the PDF that contains the survey documentation, titled "ANES 2020 Time Series Study Full Release: User Guide and Codebook." Do not be daunted by the 796-page PDF. Below, we focus on the most critical information.

Introduction

The first section in the User Guide explains that the ANES 2020 Times Series Study continues a series of election surveys conducted since 1948. These surveys contain data on public opinion and voting behavior in the U.S. presidential elections. The introduction also includes information about the modes used for data collection (web, live video interviewing, or CATI). Additionally, there is a summary of the number of pre-election interviews (8,280) and post-election re-interviews (7,449).

Sample design and respondent recruitment

The section "Sample Design and Respondent Recruitment" provides more detail about the survey's sequential mixed-mode design. All three modes were conducted one after another and not at the same time. Additionally, it indicates that for the 2020 survey, they resampled all respondents who participated in the 2016 ANES, along with a newly drawn cross-section:

> The target population for the fresh cross-section was the 231 million non-institutional U.S. citizens aged 18 or older living in the 50 U.S. states or the District of Columbia.

The document continues with more details on the sample groups.

Data analysis, weights, and variance estimation

The section "Data Analysis, Weights, and Variance Estimation" includes information on weights and strata/cluster variables. Reading through, we can find the full sample weight variables:

---

[1]https://electionstudies.org/data-center/2020-time-series-study/

> For analysis of the complete set of cases using pre-election data only, including all cases and representative of the 2020 electorate, use the full sample pre-election weight, **V200010a**. For analysis including post-election data for the complete set of participants (i.e., analysis of post-election data only or a combination of pre- and post-election data), use the full sample post-election weight, **V200010b**. Additional weights are provided for analysis of subsets of the data...

The document provides more information about the design variables, summarized in Table 3.1.

TABLE 3.1 Weight and variance information for ANES

For weight	Variance unit/cluster	Variance stratum
V200010a	V200010c	V200010d
V200010b	V200010c	V200010d

Methodology

The user guide mentions a supplemental document called "How to Analyze ANES Survey Data" (DeBell, 2010) as a how-to guide for analyzing the data. In this document, we learn more about the weights, and that they sum to the sample size and not the population. If our goal is to calculate estimates for the entire U.S. population instead of just the sample, we must adjust the weights to the U.S. population. To create accurate weights for the population, we need to determine the total population size at the time of the survey. Let's review the "Sample Design and Respondent Recruitment" section for more details:

> The target population for the fresh cross-section was the 231 million non-institutional U.S. citizens aged 18 or older living in the 50 U.S. states or the District of Columbia.

The documentation suggests that the population should equal around 231 million, but this is a very imprecise count. Upon further investigation of the available resources, we can find the methodology file titled "Methodology Report for the ANES 2020 Time Series Study" (DeBell et al., 2022). This file states that we can use the population total from the Current Population Survey (CPS), a monthly survey sponsored by the U.S. Census Bureau and the U.S. Bureau of Labor Statistics. The CPS provides a more accurate population estimate for a specific month. Therefore, we can use the CPS to get the total population number for March 2020, when the ANES was conducted. Chapter 4 goes into detailed instructions on how to calculate and adjust this value in the data.

Part II

Analysis

4

Getting started

4.1 Introduction

This chapter provides an overview of the packages, data, and design objects we use frequently throughout this book. As mentioned in Chapter 2, understanding how a survey was conducted helps us make sense of the results and interpret findings. Therefore, we provide background on the datasets used in examples and exercises. Next, we walk through how to create the survey design objects necessary to begin an analysis. Finally, we provide an overview of the {srvyr} package and the steps needed for analysis. Please report any bugs and issues encountered while going through the book to the book's GitHub repository[1].

4.2 Setup

This section provides details on the required packages and data, as well as the steps for preparing survey design objects. For a streamlined learning experience, we recommend taking the time to walk through the code provided here and making sure everything is properly set up.

4.2.1 Packages

We use several packages throughout the book, but let's install and load specific ones for this chapter. Many functions in the examples and exercises are from three packages: {tidyverse}, {survey}, and {srvyr} (Wickham et al., 2019; Lumley, 2010; Freedman Ellis and Schneider, 2024). The packages can be installed from the Comprehensive R Archive Network (CRAN) using the code below:

```
install.packages(c("tidyverse", "survey", "srvyr"))
```

---

[1]https://github.com/tidy-survey-r/tidy-survey-book

We bundled the datasets used in the book in an R package, {srvyrexploR} (Zimmer et al., 2024). To install it from GitHub, use the {pak} package (Csárdi and Hester, 2024):

```
install.packages("pak")
pak::pak("tidy-survey-r/srvyrexploR")
```

After installing these packages, load them using the library() function:

```
library(tidyverse)
library(survey)
library(srvyr)
library(srvyrexploR)
```

The packages {broom}, {gt}, and {gtsummary} play a role in displaying output and creating formatted tables (Iannone et al., 2024; Robinson et al., 2023; Sjoberg et al., 2021). Install them with the provided code[2]:

```
install.packages(c("gt", "gtsummary"))
```

After installing these packages, load them using the library() function:

```
library(broom)
library(gt)
library(gtsummary)
```

Install and load the {censusapi} package to access the Current Population Survey (CPS), which we use to ensure accurate weighting of a key dataset in the book (Recht, 2024). Run the code below to install {censusapi}:

```
install.packages("censusapi")
```

After installing this package, load it using the library() function:

```
library(censusapi)
```

Note that the {censusapi} package requires a Census API key, available for free from the U.S. Census Bureau website[3] (refer to the package documentation for more information). We recommend storing the Census API key in the R

---

[2]Note: {broom} is already included in the tidyverse, so no separate installation is required.
[3]https://api.census.gov/data/key_signup.html

environment instead of directly in the code. To do this, run the `Sys.setenv()` script below, substituting the API key where it says `YOUR_API_KEY_HERE`.

```
Sys.setenv(CENSUS_KEY = "YOUR_API_KEY_HERE")
```

Then, restart the R session. Once the Census API key is stored, we can retrieve it in our R code with `Sys.getenv("CENSUS_KEY")`.

There are a few other packages used in the book in limited frequency. We list them in the Prerequisite boxes at the beginning of each chapter. As we work through the book, make sure to check the Prerequisite box and install any missing packages before proceeding.

4.2.2 Data

The {srvyrexploR} package contains the datasets used in the book. Once installed and loaded, explore the documentation using the `help()` function. Read the descriptions of the datasets to understand what they contain:

```
help(package = "srvyrexploR")
```

This book uses two main datasets: the American National Election Studies (ANES – DeBell, 2010) and the Residential Energy Consumption Survey (RECS – U.S. Energy Information Administration, 2023b), which are included as `anes_2020` and `recs_2020` in the {srvyrexploR} package, respectively.

American National Election Studies Data

American National Election Studies (ANES) collect data from election surveys dating back to 1948. These surveys contain information on public opinion and voting behavior in U.S. presidential elections and some midterm elections[4]. They cover topics such as party affiliation, voting choice, and level of trust in the government. The 2020 survey (data used in this book) was fielded online, through live video interviews, or via computer-assisted telephone interviews (CATI).

When working with new survey data, we should review the survey documentation (see Chapter 3) to understand the data collection methods. The original ANES data contains variables starting with `v20` (DeBell, 2010), so to assist with our analysis throughout the book, we created descriptive variable names. For example, the respondent's age is now in a variable called `Age`, and gender is in a variable called `Gender`. These descriptive variables are included in the

---

[4]In the United States, presidential elections are held in years divisible by four. In other even years, there are elections at the federal level for Congress, which are referred to as midterm elections as they occur at the middle of the term of a president.

{srvyrexploR} package. A complete overview of all variables can be found in the online Appendix (Appendix B).

Before beginning an analysis, it is useful to view the data to understand the available variables. The `dplyr::glimpse()` function produces a list of all variables, their types (e.g., function, double), and a few example values. Below, we remove variables containing a "V" followed by numbers with `select(-matches("^V\\d"))` before using `glimpse()` to get a quick overview of the data with descriptive variable names:

```
anes_2020 %>%
  select(-matches("^V\\d")) %>%
  glimpse()
```

```
## Rows: 7,453
## Columns: 21
## $ CaseID                 <dbl> 200015, 200022, 200039, 200046, 2000~
## $ InterviewMode          <fct> Web, Web, Web, Web, Web, Web, Web, W~
## $ Weight                 <dbl> 1.0057, 1.1635, 0.7687, 0.5210, 0.96~
## $ VarUnit                <fct> 2, 2, 1, 2, 1, 2, 1, 2, 2, 2, 1, 1, ~
## $ Stratum                <fct> 9, 26, 41, 29, 23, 37, 7, 37, 32, 41~
## $ CampaignInterest       <fct> Somewhat interested, Not much intere~
## $ EarlyVote2020          <fct> NA, NA, NA, NA, NA, NA, NA, NA, Yes,~
## $ VotedPres2016          <fct> Yes, Yes, Yes, Yes, Yes, No, Yes, No~
## $ VotedPres2016_selection <fct> Trump, Other, Clinton, Clinton, Trum~
## $ PartyID                <fct> Strong republican, Independent, Inde~
## $ TrustGovernment        <fct> Never, Never, Some of the time, Abou~
## $ TrustPeople            <fct> About half the time, Some of the tim~
## $ Age                    <dbl> 46, 37, 40, 41, 72, 71, 37, 45, 70, ~
## $ AgeGroup               <fct> 40-49, 30-39, 40-49, 40-49, 70 or ol~
## $ Education              <fct> Bachelor's, Post HS, High school, Po~
## $ RaceEth                <fct> "Hispanic", "Asian, NH/PI", "White",~
## $ Gender                 <fct> Male, Female, Female, Male, Male, Fe~
## $ Income                 <fct> "$175,000-249,999", "$70,000-74,999"~
## $ Income7                <fct> $125k or more, $60k to < 80k, $100k ~
## $ VotedPres2020          <fct> NA, Yes, Yes, Yes, Yes, Yes, Yes, NA~
## $ VotedPres2020_selection <fct> NA, Other, Biden, Biden, Trump, Bide~
```

From the output, we can see there are 7,453 rows and 21 variables in the ANES data. This output also indicates that most of the variables are factors (e.g., `InterviewMode`), while a few variables are in double (numeric) format (e.g., `Age`).

Residential Energy Consumption Survey Data

Residential Energy Consumption Survey (RECS) is a study that measures energy consumption and expenditure in American households. Funded by the Energy Information Administration, RECS data are collected through interviews with household members and energy suppliers. These interviews take place in person, over the phone, via mail, and on the web, with modes changing over time. The survey has been fielded 14 times between 1950 and 2020. It includes questions about appliances, electronics, heating, air conditioning (A/C), temperatures, water heating, lighting, energy bills, respondent demographics, and energy assistance.

We should read the survey documentation (see Chapter 3) to understand how the data were collected and implemented. An overview of all variables can be found in the online Appendix (Appendix C).

Before starting an analysis, we recommend viewing the data to understand the types of data and variables that are included. The dplyr::glimpse() function produces a list of all variables, the type of the variable (e.g., function, double), and a few example values. Below, we remove the weight variables with select(-matches("^NWEIGHT")) before using glimpse() to get a quick overview of the data:

```
recs_2020 %>%
  select(-matches("^NWEIGHT")) %>%
  glimpse()
```

```
## Rows: 18,496
## Columns: 39
## $ DOEID           <dbl> 1e+05, 1e+05, 1e+05, 1e+05, 1e+05, 1e+05, 1~
## $ ClimateRegion_BA <fct> Mixed-Dry, Mixed-Humid, Mixed-Dry, Mixed-Hu~
## $ Urbanicity      <fct> Urban Area, Urban Area, Urban Area, Urban A~
## $ Region          <fct> West, South, West, South, Northeast, South,~
## $ REGIONC         <chr> "WEST", "SOUTH", "WEST", "SOUTH", "NORTHEAS~
## $ Division        <fct> Mountain South, West South Central, Mountai~
## $ STATE_FIPS      <chr> "35", "05", "35", "45", "34", "48", "40", "~
## $ state_postal    <fct> NM, AR, NM, SC, NJ, TX, OK, MS, DC, AZ, CA,~
## $ state_name      <fct> New Mexico, Arkansas, New Mexico, South Car~
## $ HDD65           <dbl> 3844, 3766, 3819, 2614, 4219, 901, 3148, 18~
## $ CDD65           <dbl> 1679, 1458, 1696, 1718, 1363, 3558, 2128, 2~
## $ HDD30YR         <dbl> 4451, 4429, 4500, 3229, 4896, 1150, 3564, 2~
## $ CDD30YR         <dbl> 1027, 1305, 1010, 1653, 1059, 3588, 2043, 2~
## $ HousingUnitType <fct> Single-family detached, Apartment: 5 or mor~
## $ YearMade        <ord> 1970-1979, 1980-1989, 1960-1969, 1980-1989,~
## $ TOTSQFT_EN      <dbl> 2100, 590, 900, 2100, 800, 4520, 2100, 900,~
## $ TOTHSQFT        <dbl> 2100, 590, 900, 2100, 800, 3010, 1200, 900,~
## $ TOTCSQFT        <dbl> 2100, 590, 900, 2100, 800, 3010, 1200, 0, 5~
## $ SpaceHeatingUsed <lgl> TRUE, TRUE, TRUE, TRUE, TRUE, TRUE, TRUE, T~
## $ ACUsed          <lgl> TRUE, TRUE, TRUE, TRUE, TRUE, TRUE, TRUE, F~
## $ HeatingBehavior <fct> Set one temp and leave it, Turn on or off a~
## $ WinterTempDay   <dbl> 70, 70, 69, 68, 68, 76, 74, 70, 68, 70, 72,~
## $ WinterTempAway  <dbl> 70, 65, 68, 68, 68, 76, 65, 70, 60, 70, 70,~
## $ WinterTempNight <dbl> 68, 65, 67, 68, 68, 68, 74, 68, 62, 68, 72,~
```

```
## $ ACBehavior       <fct> Set one temp and leave it, Turn on or off a~
## $ SummerTempDay    <dbl> 71, 68, 70, 72, 72, 69, 68, NA, 72, 74, 77,~
## $ SummerTempAway   <dbl> 71, 68, 68, 72, 72, 74, 70, NA, 76, 74, 77,~
## $ SummerTempNight  <dbl> 71, 68, 68, 72, 72, 68, 70, NA, 68, 72, 77,~
## $ BTUEL            <dbl> 42723, 17889, 8147, 31647, 20027, 48968, 49~
## $ DOLLAREL         <dbl> 1955.06, 713.27, 334.51, 1424.86, 1087.00, ~
## $ BTUNG            <dbl> 101924.4, 10145.3, 22603.1, 55118.7, 39099.~
## $ DOLLARNG         <dbl> 701.83, 261.73, 188.14, 636.91, 376.04, 439~
## $ BTULP            <dbl> 0, 0, 0, 0, 0, 0, 0, 0, 0, 0, 0, 0, 0, 0, 1~
## $ DOLLARLP         <dbl> 0.0, 0.0, 0.0, 0.0, 0.0, 0.0, 0.0, 0.0, 0.0~
## $ BTUFO            <dbl> 0, 0, 0, 0, 0, 0, 0, 0, 0, 0, 0, 0, 0, 0, 6~
## $ DOLLARFO         <dbl> 0, 0, 0, 0, 0, 0, 0, 0, 0, 0, 0, 0, 0, 0, 1~
## $ BTUWOOD          <dbl> 0, 0, 0, 0, 0, 3000, 0, 0, 0, 0, 0, 0, 0, 0~
## $ TOTALBTU         <dbl> 144648, 28035, 30750, 86765, 59127, 85401, ~
## $ TOTALDOL         <dbl> 2656.9, 975.0, 522.6, 2061.8, 1463.0, 2335.~
```

From the output, we can see that the RECS data has 18,496 rows and 39 non-weight variables. This output also indicates that most of the variables are in double (numeric) format (e.g., TOTSQFT_EN), with some factor (e.g., Region), Boolean (e.g., ACUsed), character (e.g., REGIONC), and ordinal (e.g., YearMade) variables.

4.2.3   Design objects

The design object is the backbone for survey analysis. It is where we specify the sampling design, weights, and other necessary information to ensure we account for errors in the data. Before creating the design object, we should carefully review the survey documentation to understand how to create the design object for accurate analysis.

In this section, we provide details on how to code the design object for the ANES and RECS data used in the book. However, we only provide a high-level overview to get readers started. For a deeper understanding of creating design objects for a variety of sampling designs, see Chapter 10.

While we recommend conducting exploratory data analysis on the original data before diving into complex survey analysis (see Chapter 12), the actual survey analysis and inference should be performed with the survey design objects instead of the original survey data. For example, the ANES data is called anes_2020. If we create a survey design object called anes_des, our survey analyses should begin with anes_des and not anes_2020. Using the survey design object ensures that our calculations appropriately account for the details of the survey design.

American National Election Studies Design Object

The ANES documentation (DeBell, 2010) details the sampling and weighting implications for analyzing the survey data. From this documentation and as noted in Chapter 3, the 2020 ANES data are weighted to the sample, not the population. To make generalizations about the population, we need to weigh the data against the full population count. The ANES methodology recommends using the Current Population Survey (CPS) to determine the number of non-institutional U.S. citizens aged 18 or older living in the 50 U.S. states or D.C. in March 2020.

We can use the {censusapi} package to obtain the information needed for the survey design object. The getCensus() function allows us to retrieve the CPS data for March (cps/basic/mar) in 2020 (vintage = 2020). Additionally, we extract several variables from the CPS:

- month (HRMONTH) and year (HRYEAR4) of the interview: to confirm the correct time period
- age (PRTAGE) of the respondent: to narrow the population to 18 and older (eligible age to vote)
- citizenship status (PRCITSHP) of the respondent: to narrow the population to only those eligible to vote
- final person-level weight (PWSSWGT)

Detailed information for these variables can be found in the CPS data dictionary[5].

```
cps_state_in <- getCensus(
  name = "cps/basic/mar",
  vintage = 2020,
  region = "state",
  vars = c(
    "HRMONTH", "HRYEAR4",
    "PRTAGE", "PRCITSHP", "PWSSWGT"
  ),
  key = Sys.getenv("CENSUS_KEY")
)


cps_state <- cps_state_in %>%
  as_tibble() %>%
  mutate(across(
    .cols = everything(),
    .fns = as.numeric
  ))
```

---

[5]https://www2.census.gov/programs-surveys/cps/datasets/2020/basic/2020_Basic_CPS_Public_Use_Record_Layout_plus_IO_Code_list.txt

In the code above, we include `region = "state"`. The default region type for the CPS data is at the state level. While not required, including the region can be helpful for understanding the geographical context of the data.

In `getCensus()`, we filtered the dataset by specifying the month (`HRMONTH ==` `3`) and year (`HRYEAR4 == 2020`) of our request. Therefore, we expect that all interviews within our output were conducted during that particular month and year. We can confirm that the data are from March 2020 by running the code below:

```
cps_state %>%
  distinct(HRMONTH, HRYEAR4)
```

```
## # A tibble: 1 x 2
##   HRMONTH HRYEAR4
##     <dbl>   <dbl>
## 1       3    2020
```

We can narrow down the dataset using the age and citizenship variables to include only individuals who are 18 years or older (`PRTAGE >= 18`) and have U.S. citizenship (`PRCITSHIP %in% c(1:4)`):

```
cps_narrow_resp <- cps_state %>%
  filter(
    PRTAGE >= 18,
    PRCITSHP %in% c(1:4)
  )
```

To calculate the U.S. population from the filtered data, we sum the person weights (`PWSSWGT`):

```
targetpop <- cps_narrow_resp %>%
  pull(PWSSWGT) %>%
  sum()
```

```
scales::comma(targetpop)
```

```
## [1] "231,034,125"
```

The population of interest in 2020 is 231,034,125. This result gives us what we need to create the survey design object for estimating population statistics. Using the anes_2020 data, we adjust the weighting variable (`V200010b`) using the population of interest we just calculated (`targetpop`). We determine the proportion of the total weight for each individual weight (`V200010b / sum(V200010b)`) and then multiply that proportion by the calculated population of interest.

```
anes_adjwgt <- anes_2020 %>%
  mutate(Weight = V200010b / sum(V200010b) * targetpop)
```

Once we have the adjusted weights, we can refer to the rest of the documentation to create the survey design. The documentation indicates that the study uses a stratified cluster sampling design. Therefore, we need to specify variables for strata and ids (cluster) and fill in the nest argument. The documentation provides guidance on which strata and cluster variables to use depending on whether we are analyzing pre- or post-election data. In this book, we analyze post-election data, so we need to use the post-election weight V200010b, strata variable V200010d, and Primary Sampling Unit (PSU)/cluster variable V200010c. Additionally, we set nest=TRUE to ensure the clusters are nested within the strata.

```
anes_des <- anes_adjwgt %>%
  as_survey_design(
    weights = Weight,
    strata = V200010d,
    ids = V200010c,
    nest = TRUE
  )

anes_des
```

```
## Stratified 1 - level Cluster Sampling design (with replacement)
## With (101) clusters.
## Called via srvyr
## Sampling variables:
##    - ids: V200010c
##    - strata: V200010d
##    - weights: Weight
## Data variables:
##    - V200001 (dbl), CaseID (dbl), V200002 (dbl+lbl), InterviewMode
##      (fct), V200010b (dbl), Weight (dbl), V200010c (dbl), VarUnit
##      (fct), V200010d (dbl), Stratum (fct), V201006 (dbl+lbl),
##      CampaignInterest (fct), V201023 (dbl+lbl), EarlyVote2020 (fct),
##      V201024 (dbl+lbl), V201025x (dbl+lbl), V201028 (dbl+lbl), V201029
##      (dbl+lbl), V201101 (dbl+lbl), V201102 (dbl+lbl), VotedPres2016
##      (fct), V201103 (dbl+lbl), VotedPres2016_selection (fct), V201228
##      (dbl+lbl), V201229 (dbl+lbl), V201230 (dbl+lbl), V201231x
##      (dbl+lbl), PartyID (fct), V201233 (dbl+lbl), TrustGovernment
##      (fct), V201237 (dbl+lbl), TrustPeople (fct), V201507x (dbl+lbl),
##      Age (dbl), AgeGroup (fct), V201510 (dbl+lbl), Education (fct),
##      V201546 (dbl+lbl), V201547a (dbl+lbl), V201547b (dbl+lbl),
##      V201547c (dbl+lbl), V201547d (dbl+lbl), V201547e (dbl+lbl),
##      V201547z (dbl+lbl), V201549x (dbl+lbl), RaceEth (fct), V201600
```

```
##      (dbl+lbl), Gender (fct), V201607 (dbl+lbl), V201610 (dbl+lbl),
##      V201611 (dbl+lbl), V201613 (dbl+lbl), V201615 (dbl+lbl), V201616
##      (dbl+lbl), V201617x (dbl+lbl), Income (fct), Income7 (fct),
##      V202051 (dbl+lbl), V202066 (dbl+lbl), V202072 (dbl+lbl),
##      VotedPres2020 (fct), V202073 (dbl+lbl), V202109x (dbl+lbl),
##      V202110x (dbl+lbl), VotedPres2020_selection (fct)
```

We can examine this new object to learn more about the survey design, such that the ANES is a "Stratified 1 - level Cluster Sampling design (with replacement) With (101) clusters." Additionally, the output displays the sampling variables and then lists the remaining variables in the dataset. This design object is used throughout this book to conduct survey analysis.

Residential Energy Consumption Survey Design Object

The RECS documentation (U.S. Energy Information Administration, 2023b) provides information on the survey's sampling and weighting implications for analysis. The documentation shows the 2020 RECS uses Jackknife weights, where the main analytic weight is NWEIGHT, and the Jackknife weights are NWEIGHT1-NWEIGHT60. We can specify these in the weights and repweights arguments in the survey design object code, respectively.

With Jackknife weights, additional information is required: type, scale, and mse. Chapter 10 discusses in depth each of these arguments; but to quickly get started, the RECS documentation lets us know that type=JK1, scale=59/60, and mse = TRUE. We can use the following code to create the survey design object:

```
recs_des <- recs_2020 %>%
  as_survey_rep(
    weights = NWEIGHT,
    repweights = NWEIGHT1:NWEIGHT60,
    type = "JK1",
    scale = 59 / 60,
    mse = TRUE
  )

recs_des
```

```
## Call: Called via srvyr
## Unstratified cluster jacknife (JK1) with 60 replicates and MSE variances.
## Sampling variables:
##  - repweights: `NWEIGHT1 + NWEIGHT2 + NWEIGHT3 + NWEIGHT4 + NWEIGHT5
##    + NWEIGHT6 + NWEIGHT7 + NWEIGHT8 + NWEIGHT9 + NWEIGHT10 +
##    NWEIGHT11 + NWEIGHT12 + NWEIGHT13 + NWEIGHT14 + NWEIGHT15 +
##    NWEIGHT16 + NWEIGHT17 + NWEIGHT18 + NWEIGHT19 + NWEIGHT20 +
##    NWEIGHT21 + NWEIGHT22 + NWEIGHT23 + NWEIGHT24 + NWEIGHT25 +
```

```
##        NWEIGHT26 + NWEIGHT27 + NWEIGHT28 + NWEIGHT29 + NWEIGHT30 +
##        NWEIGHT31 + NWEIGHT32 + NWEIGHT33 + NWEIGHT34 + NWEIGHT35 +
##        NWEIGHT36 + NWEIGHT37 + NWEIGHT38 + NWEIGHT39 + NWEIGHT40 +
##        NWEIGHT41 + NWEIGHT42 + NWEIGHT43 + NWEIGHT44 + NWEIGHT45 +
##        NWEIGHT46 + NWEIGHT47 + NWEIGHT48 + NWEIGHT49 + NWEIGHT50 +
##        NWEIGHT51 + NWEIGHT52 + NWEIGHT53 + NWEIGHT54 + NWEIGHT55 +
##        NWEIGHT56 + NWEIGHT57 + NWEIGHT58 + NWEIGHT59 + NWEIGHT60`
##    - weights: NWEIGHT
## Data variables:
##    - DOEID (dbl), ClimateRegion_BA (fct), Urbanicity (fct), Region
##      (fct), REGIONC (chr), Division (fct), STATE_FIPS (chr),
##      state_postal (fct), state_name (fct), HDD65 (dbl), CDD65 (dbl),
##      HDD30YR (dbl), CDD30YR (dbl), HousingUnitType (fct), YearMade
##      (ord), TOTSQFT_EN (dbl), TOTHSQFT (dbl), TOTCSQFT (dbl),
##      SpaceHeatingUsed (lgl), ACUsed (lgl), HeatingBehavior (fct),
##      WinterTempDay (dbl), WinterTempAway (dbl), WinterTempNight (dbl),
##      ACBehavior (fct), SummerTempDay (dbl), SummerTempAway (dbl),
##      SummerTempNight (dbl), NWEIGHT (dbl), NWEIGHT1 (dbl), NWEIGHT2
##      (dbl), NWEIGHT3 (dbl), NWEIGHT4 (dbl), NWEIGHT5 (dbl), NWEIGHT6
##      (dbl), NWEIGHT7 (dbl), NWEIGHT8 (dbl), NWEIGHT9 (dbl), NWEIGHT10
##      (dbl), NWEIGHT11 (dbl), NWEIGHT12 (dbl), NWEIGHT13 (dbl),
##      NWEIGHT14 (dbl), NWEIGHT15 (dbl), NWEIGHT16 (dbl), NWEIGHT17
##      (dbl), NWEIGHT18 (dbl), NWEIGHT19 (dbl), NWEIGHT20 (dbl),
##      NWEIGHT21 (dbl), NWEIGHT22 (dbl), NWEIGHT23 (dbl), NWEIGHT24
##      (dbl), NWEIGHT25 (dbl), NWEIGHT26 (dbl), NWEIGHT27 (dbl),
##      NWEIGHT28 (dbl), NWEIGHT29 (dbl), NWEIGHT30 (dbl), NWEIGHT31
##      (dbl), NWEIGHT32 (dbl), NWEIGHT33 (dbl), NWEIGHT34 (dbl),
##      NWEIGHT35 (dbl), NWEIGHT36 (dbl), NWEIGHT37 (dbl), NWEIGHT38
##      (dbl), NWEIGHT39 (dbl), NWEIGHT40 (dbl), NWEIGHT41 (dbl),
##      NWEIGHT42 (dbl), NWEIGHT43 (dbl), NWEIGHT44 (dbl), NWEIGHT45
##      (dbl), NWEIGHT46 (dbl), NWEIGHT47 (dbl), NWEIGHT48 (dbl),
##      NWEIGHT49 (dbl), NWEIGHT50 (dbl), NWEIGHT51 (dbl), NWEIGHT52
##      (dbl), NWEIGHT53 (dbl), NWEIGHT54 (dbl), NWEIGHT55 (dbl),
##      NWEIGHT56 (dbl), NWEIGHT57 (dbl), NWEIGHT58 (dbl), NWEIGHT59
##      (dbl), NWEIGHT60 (dbl), BTUEL (dbl), DOLLAREL (dbl), BTUNG (dbl),
##      DOLLARNG (dbl), BTULP (dbl), DOLLARLP (dbl), BTUFO (dbl),
##      DOLLARFO (dbl), BTUWOOD (dbl), TOTALBTU (dbl), TOTALDOL (dbl)
```

Viewing this new object provides information about the survey design, such that RECS is an "Unstratified cluster jacknife (JK1) with 60 replicates and MSE variances." Additionally, the output shows the sampling variables (NWEIGHT1-NWEIGHT60) and then lists the remaining variables in the dataset. This design object is used throughout this book to conduct survey analysis.

4.3   Survey analysis process

There is a general process for analyzing data to create estimates with {srvyr} package:

1.   Create a `tbl_svy` object (a survey object) using: `as_survey_design()` or `as_survey_rep()`

2.   Subset data (if needed) using `filter()` (to create subpopulations)

3.   Specify domains of analysis using `group_by()`

4.   Within `summarize()`, specify variables to calculate, including means, totals, proportions, quantiles, and more

In Section 4.2.3, we follow Step 1 to create the survey design objects for the ANES and RECS data featured in this book. Additional details on how to create design objects can be found in Chapter 10. Then, once we have the design object, we can filter the data to any subpopulation of interest (if needed). It is important to filter the data after creating the design object. This ensures that we are accurately accounting for the survey design in our calculations. Finally, we can use `group_by()`, `summarize()`, and other functions from the {survey} and {srvyr} packages to analyze the survey data by estimating means, totals, and so on.

4.4   Similarities between {dplyr} and {srvyr} functions

The {dplyr} package from the tidyverse offers flexible and intuitive functions for data wrangling (Wickham et al., 2023a). One of the major advantages of using {srvyr} is that it applies {dplyr}-like syntax to the {survey} package (Freedman Ellis and Schneider, 2024). We can use pipes, such as `%>%` from the {magrittr} package, to specify a survey design object, apply a function, and then feed that output into the next function's first argument (Bache and Wickham, 2022). Functions follow the 'tidy' convention of snake_case function names.

To help explain the similarities between {dplyr} functions and {srvyr} functions, we use the `towny` dataset from the {gt} package and `apistrat` data that comes in the {survey} package. The `towny` dataset provides population data for municipalities in Ontario, Canada on census years between 1996 and 2021. Taking a look at `towny` with `dplyr::glimpse()`, we can see the dataset has 25 columns with a mix of character and numeric data.

```
towny %>%
  glimpse()
```

```
## Rows: 414
## Columns: 25
## $ name                      <chr> "Addington Highlands", "Adelaide Me~
## $ website                   <chr> "https://addingtonhighlands.ca", "h~
## $ status                    <chr> "lower-tier", "lower-tier", "lower-~
## $ csd_type                  <chr> "township", "township", "township",~
## $ census_div                <chr> "Lennox and Addington", "Middlesex"~
## $ latitude                  <dbl> 45.00, 42.95, 44.13, 45.53, 43.86, ~
## $ longitude                 <dbl> -77.25, -81.70, -79.93, -76.90, -79~
## $ land_area_km2             <dbl> 1293.99, 331.11, 371.53, 519.59, 66~
## $ population_1996           <int> 2429, 3128, 9359, 2837, 64430, 1027~
## $ population_2001           <int> 2402, 3149, 10082, 2824, 73753, 956~
## $ population_2006           <int> 2512, 3135, 10695, 2716, 90167, 958~
## $ population_2011           <int> 2517, 3028, 10603, 2844, 109600, 86~
## $ population_2016           <int> 2318, 2990, 10975, 2935, 119677, 96~
## $ population_2021           <int> 2534, 3011, 10989, 2995, 126666, 95~
## $ density_1996              <dbl> 1.88, 9.45, 25.19, 5.46, 966.84, 8.~
## $ density_2001              <dbl> 1.86, 9.51, 27.14, 5.44, 1106.74, 8~
## $ density_2006              <dbl> 1.94, 9.47, 28.79, 5.23, 1353.05, 8~
## $ density_2011              <dbl> 1.95, 9.14, 28.54, 5.47, 1644.66, 7~
## $ density_2016              <dbl> 1.79, 9.03, 29.54, 5.65, 1795.87, 8~
## $ density_2021              <dbl> 1.96, 9.09, 29.58, 5.76, 1900.75, 8~
## $ pop_change_1996_2001_pct  <dbl> -0.0111, 0.0067, 0.0773, -0.0046, 0~
## $ pop_change_2001_2006_pct  <dbl> 0.0458, -0.0044, 0.0608, -0.0382, 0~
## $ pop_change_2006_2011_pct  <dbl> 0.0020, -0.0341, -0.0086, 0.0471, 0~
## $ pop_change_2011_2016_pct  <dbl> -0.0791, -0.0125, 0.0351, 0.0320, 0~
## $ pop_change_2016_2021_pct  <dbl> 0.0932, 0.0070, 0.0013, 0.0204, 0.0~
```

Let's examine the `towny` object's class. We verify that it is a tibble, as indicated by `"tbl_df"`, by running the code below:

```
class(towny)
```

```
## [1] "tbl_df"      "tbl"         "data.frame"
```

All tibbles are data.frames, but not all data.frames are tibbles. Compared to data.frames, tibbles have some advantages, with the printing behavior being a noticeable advantage. When working with tidyverse style code, we recommend making all your datasets tibbles for ease of analysis.

The {survey} package contains datasets related to the California Academic Performance Index, which measures student performance in schools with at least 100 students in California. We can access these datasets by loading the {survey} package and running `data(api)`.

Let's work with the `apistrat` dataset, which is a stratified random sample, stratified by school type (`stype`) with three levels: `E` for elementary school, `M` for middle school, and `H` for high school. We first create the survey design object (see Chapter 10 for more information). The sample is stratified by the `stype` variable and the sampling weights are found in the `pw` variable. We can use this information to construct the design object, `apistrat_des`.

```
data(api)

apistrat_des <- apistrat %>%
  as_survey_design(
    strata = stype,
    weights = pw
  )
```

When we check the class of `apistrat_des`, it is not a typical `data.frame`. Applying the `as_survey_design()` function transforms the data into a `tbl_svy`, a special class specifically for survey design objects. The {srvyr} package is designed to work with the `tbl_svy` class of objects.

```
class(apistrat_des)
```

```
## [1] "tbl_svy"         "survey.design2" "survey.design"
```

Let's look at how {dplyr} works with regular data frames. The example below calculates the mean and median for the `land_area_km2` variable in the `towny` dataset.

```
towny %>%
  summarize(
    area_mean = mean(land_area_km2),
    area_median = median(land_area_km2)
  )
```

```
## # A tibble: 1 x 2
##   area_mean area_median
##       <dbl>       <dbl>
## 1      373.        273.
```

In the code below, we calculate the mean and median of the variable `api00` using `apistrat_des`. Note the similarity in the syntax. However, the standard error of the statistic is also calculated in addition to the statistic itself.

```
apistrat_des %>%
  summarize(
    api00_mean = survey_mean(api00),
    api00_med = survey_median(api00)
  )
```

```
## # A tibble: 1 x 4
##   api00_mean api00_mean_se api00_med api00_med_se
##        <dbl>         <dbl>     <dbl>        <dbl>
## 1       662.          9.54       668         13.7
```

The functions in {srvyr} also play nicely with other tidyverse functions. For example, if we wanted to select columns with shared characteristics, we can use {tidyselect} functions such as starts_with(), num_range(), etc. (Henry and Wickham, 2024). In the examples below, we use a combination of across() and starts_with() to calculate the mean of variables starting with "population" in the towny data frame and those beginning with api in the apistrat_des survey object.

```
towny %>%
  summarize(across(
    starts_with("population"),
    ~ mean(.x, na.rm = TRUE)
  ))
```

```
## # A tibble: 1 x 6
##   population_1996 population_2001 population_2006 population_2011
##             <dbl>           <dbl>           <dbl>           <dbl>
## 1          25866.          27538.          29173.          30838.
## # i 2 more variables: population_2016 <dbl>, population_2021 <dbl>
```

```
apistrat_des %>%
  summarize(across(
    starts_with("api"),
    survey_mean
  ))
```

```
## # A tibble: 1 x 6
##   api00 api00_se api99 api99_se api.stu api.stu_se
##   <dbl>    <dbl> <dbl>    <dbl>   <dbl>      <dbl>
## 1  662.     9.54  629.     10.1    498.       16.4
```

We have the flexibility to use {dplyr} verbs such as mutate(), filter(), and
select() on our survey design object. As mentioned in Section 4.3, these steps
should be performed on the survey design object. This ensures our survey
design is properly considered in all our calculations.

```
apistrat_des_mod <- apistrat_des %>%
  mutate(api_diff = api00 - api99) %>%
  filter(stype == "E") %>%
  select(stype, api99, api00, api_diff, api_students = api.stu)
```

```
apistrat_des_mod
```

```
## Stratified Independent Sampling design (with replacement)
## Called via srvyr
## Sampling variables:
##    - ids: `1`
##    - strata: stype
##    - weights: pw
## Data variables:
##    - stype (fct), api99 (int), api00 (int), api_diff (int),
##      api_students (int)
```

```
apistrat_des
```

```
## Stratified Independent Sampling design (with replacement)
## Called via srvyr
## Sampling variables:
##    - ids: `1`
##    - strata: stype
##    - weights: pw
## Data variables:
##    - cds (chr), stype (fct), name (chr), sname (chr), snum (dbl),
##      dname (chr), dnum (int), cname (chr), cnum (int), flag (int),
##      pcttest (int), api00 (int), api99 (int), target (int), growth
##      (int), sch.wide (fct), comp.imp (fct), both (fct), awards (fct),
##      meals (int), ell (int), yr.rnd (fct), mobility (int), acs.k3
##      (int), acs.46 (int), acs.core (int), pct.resp (int), not.hsg
##      (int), hsg (int), some.col (int), col.grad (int), grad.sch (int),
##      avg.ed (dbl), full (int), emer (int), enroll (int), api.stu
##      (int), pw (dbl), fpc (dbl)
```

Several functions in {srvyr} must be called within srvyr::summarize(), with
the exception of srvyr::survey_count() and srvyr::survey_tally(). This is
similar to how dplyr::count() and dplyr::tally() are not called within
dplyr::summarize(). The summarize() function can be used in conjunction

with the group_by() function or by/.by arguments, which applies the functions on a group-by-group basis to create grouped summaries.

```
towny %>%
  group_by(csd_type) %>%
  dplyr::summarize(
    area_mean = mean(land_area_km2),
    area_median = median(land_area_km2)
  )
```

```
## # A tibble: 5 x 3
##   csd_type      area_mean area_median
##   <chr>             <dbl>       <dbl>
## 1 city              498.        198.
## 2 municipality      607.        488.
## 3 town              183.        129.
## 4 township          363.        301.
## 5 village            23.0         3.3
```

We use a similar setup to summarize data in {srvyr}:

```
apistrat_des %>%
  group_by(stype) %>%
  summarize(
    api00_mean = survey_mean(api00),
    api00_median = survey_median(api00)
  )
```

```
## # A tibble: 3 x 5
##   stype api00_mean api00_mean_se api00_median api00_median_se
##   <fct>      <dbl>         <dbl>        <dbl>           <dbl>
## 1 E          674.          12.5          671            20.7
## 2 H          626.          15.5          635            21.6
## 3 M          637.          16.6          648            24.1
```

An alternative way to do grouped analysis on the towny data would be with the .by argument:

```
towny %>%
  dplyr::summarize(
    area_mean = mean(land_area_km2),
    area_median = median(land_area_km2),
    .by = csd_type
  )
```

```
## # A tibble: 5 x 3
##   csd_type       area_mean area_median
##   <chr>              <dbl>       <dbl>
## 1 township            363.        301.
## 2 town                183.        129.
## 3 municipality        607.        488.
## 4 city                498.        198.
## 5 village            23.0         3.3
```

The .by syntax is similarly implemented in {srvyr} for grouped analysis:

```
apistrat_des %>%
  summarize(
    api00_mean = survey_mean(api00),
    api00_median = survey_median(api00),
    .by = stype
  )
```

```
## # A tibble: 3 x 5
##   stype api00_mean api00_mean_se api00_median api00_median_se
##   <fct>      <dbl>         <dbl>        <dbl>           <dbl>
## 1 E           674.          12.5          671            20.7
## 2 H           626.          15.5          635            21.6
## 3 M           637.          16.6          648            24.1
```

As mentioned above, {srvyr} functions are meant for tbl_svy objects. Attempting to manipulate data on non-tbl_svy objects, like the towny example shown below, results in an error. Running the code lets us know what the issue is: Survey context not set.

```
towny %>%
  summarize(area_mean = survey_mean(land_area_km2))
```

```
## Error in `summarize()`:
## i In argument: `area_mean = survey_mean(land_area_km2)`.
## Caused by error in `cur_svy()`:
## ! Survey context not set
```

A few functions in {srvyr} have counterparts in {dplyr}, such as srvyr::summarize() and srvyr::group_by(). Unlike {srvyr}-specific verbs, {srvyr} recognizes these parallel functions if applied to a non-survey object. Instead of causing an error, the package provides the equivalent output from {dplyr}:

```
towny %>%
  srvyr::summarize(area_mean = mean(land_area_km2))
```

```
## # A tibble: 1 x 1
##    area_mean
##        <dbl>
## 1      373.
```

Because this book focuses on survey analysis, most of our pipes stem from a survey object. When we load the {dplyr} and {srvyr} packages, the functions automatically figure out the class of data and use the appropriate one from {dplyr} or {srvyr}. Therefore, we do not need to include the namespace for each function (e.g., `srvyr::summarize()`).

5

Descriptive analyses

We are using data from ANES and RECS described in Chapter 4. As a reminder, here is the code to create the design objects for each to use throughout this chapter. For ANES, we need to adjust the weight so it sums to the population instead of the sample (see the ANES documentation and Chapter 4 for more information).

```
targetpop <- 231592693

anes_adjwgt <- anes_2020 %>%
  mutate(Weight = Weight / sum(Weight) * targetpop)

anes_des <- anes_adjwgt %>%
  as_survey_design(
    weights = Weight,
    strata = Stratum,
    ids = VarUnit,
    nest = TRUE
  )
```

For RECS, details are included in the RECS documentation and Chapters 4 and 10.

```
recs_des <- recs_2020 %>%
  as_survey_rep(
    weights = NWEIGHT,
    repweights = NWEIGHT1:NWEIGHT60,
    type = "JK1",
    scale = 59 / 60,
    mse = TRUE
  )
```

5.1   Introduction

Descriptive analyses, such as basic counts, cross-tabulations, or means, are among the first steps in making sense of our survey results. During descriptive analyses, we calculate point estimates of unknown population parameters, such as population mean, and uncertainty estimates, such as confidence intervals. By reviewing the findings, we can glean insight into the data, the underlying population, and any unique aspects of the data or population. For example, if only 10% of survey respondents are male, it could indicate a unique population, a potential error or bias, an intentional survey sampling method, or other factors. Additionally, descriptive analyses provide summaries of distribution and other measures. These analyses lay the groundwork for the next steps of running statistical tests or developing models.

We discuss many different types of descriptive analyses in this chapter. However, it is important to know what type of data we are working with and which statistics are appropriate. In survey data, we typically consider data as one of four main types:

- Categorical/nominal data: variables with levels or descriptions that cannot be ordered, such as the region of the country (North, South, East, and West)
- Ordinal data: variables that can be ordered, such as those from a Likert scale (strongly disagree, disagree, agree, and strongly agree)
- Discrete data: variables that are counted or measured, such as number of children
- Continuous data: variables that are measured and whose values can lie anywhere on an interval, such as income

This chapter discusses how to analyze measures of distribution (e.g., cross-tabulations), central tendency (e.g., means), relationship (e.g., ratios), and dispersion (e.g., standard deviation) using functions from the {srvyr} package (Freedman Ellis and Schneider, 2024).

Measures of distribution describe how often an event or response occurs. These measures include counts and totals. We cover the following functions:

- Count of observations (`survey_count()` and `survey_tally()`)
- Summation of variables (`survey_total()`)

Measures of central tendency find the central (or average) responses. These measures include means and medians. We cover the following functions:

- Means and proportions (`survey_mean()` and `survey_prop()`)
- Quantiles and medians (`survey_quantile()` and `survey_median()`)

Measures of relationship describe how variables relate to each other. These measures include correlations and ratios. We cover the following functions:

- Correlations (`survey_corr()`)
- Ratios (`survey_ratio()`)

Measures of dispersion describe how data spread around the central tendency for continuous variables. These measures include standard deviations and variances. We cover the following functions:

- Variances and standard deviations (`survey_var()` and `survey_sd()`)

To incorporate each of these survey functions, recall the general process for survey estimation from Chapter 4:

1. Create a `tbl_svy` object using `srvyr::as_survey_design()` or `srvyr::as_survey_rep()`.
2. Subset the data for subpopulations using `srvyr::filter()`, if needed.
3. Specify domains of analysis using `srvyr::group_by()`, if needed.
4. Analyze the data with survey-specific functions.

This chapter walks through how to apply the survey functions in Step 4. Note that unless otherwise specified, our estimates are weighted as a result of setting up the survey design object.

To look at the data by different subgroups, we can choose to filter and/or group the data. It is very important that we filter and group the data only after creating the design object. This ensures that the results accurately reflect the survey design. If we filter or group data before creating the survey design object, the data for those cases are not included in the survey design information and estimations of the variance, leading to inaccurate results.

For the sake of simplicity, we've removed cases with missing values in the examples below. For a more detailed explanation of how to handle missing data, please refer to Chapter 11.

5.2   Counts and cross-tabulations

Using `survey_count()` and `survey_tally()`, we can calculate the estimated population counts for a given variable or combination of variables. These summaries, often referred to as cross-tabulations or cross-tabs, are applied to categorical data. They help in estimating counts of the population size for different groups based on the survey data.

5.2.1   Syntax

The syntax for `survey_count()` is similar to the `dplyr::count()` syntax, as mentioned in Chapter 4. However, as noted above, this function can only be called on `tbl_svy` objects. Let's explore the syntax:

```
survey_count(
  x,
  ...,
  wt = NULL,
  sort = FALSE,
  name = "n",
  .drop = dplyr::group_by_drop_default(x),
  vartype = c("se", "ci", "var", "cv")
)
```

The arguments are:

- x: a `tbl_svy` object created by `as_survey`
- ...: variables to group by, passed to `group_by`
- wt: a variable to weight on in addition to the survey weights, defaults to NULL
- sort: how to sort the variables, defaults to FALSE
- name: the name of the count variable, defaults to n
- .drop: whether to drop empty groups
- vartype: type(s) of variation estimate to calculate including any of c("se", "ci", "var", "cv"), defaults to se (standard error) (see Section 5.2.1 for more information)

To generate a count or cross-tabs by different variables, we include them in the (...) argument. This argument can take any number of variables and breaks down the counts by all combinations of the provided variables. This is similar to `dplyr::count()`. To obtain an estimate of the overall population, we can exclude any variables from the (...) argument or use the `survey_tally()` function. While the `survey_tally()` function has a similar syntax to the `survey_count()` function, it does not include the (...) or the .drop arguments:

```
survey_tally(
  x,
  wt,
  sort = FALSE,
  name = "n",
  vartype = c("se", "ci", "var", "cv")
)
```

Both functions include the vartype argument with four different values:

- se: standard error
  - The estimated standard deviation of the estimate
  - Output has a column with the variable name specified in the name argument with a suffix of "_se"
- ci: confidence interval
  - The lower and upper limits of a confidence interval
  - Output has two columns with the variable name specified in the name argument with a suffix of "_low" and "_upp"
  - By default, this is a 95% confidence interval but can be changed by using the argument level and specifying a number between 0 and 1. For example, level=0.8 would produce an 80% confidence interval.
- var: variance
  - The estimated variance of the estimate
  - Output has a column with the variable name specified in the name argument with a suffix of "_var"
- cv: coefficient of variation
  - A ratio of the standard error and the estimate
  - Output has a column with the variable name specified in the name argument with a suffix of "_cv"

The confidence intervals are always calculated using a symmetric t-distribution based method, given by the formula:

$$\text{estimate} \pm t^*_{df} \times SE$$

where t^*_{df} is the critical value from a t-distribution based on the confidence level and the degrees of freedom. By default, the degrees of freedom are based on the design or number of replicates, but they can be specified using the df argument. For survey design objects, the degrees of freedom are calculated as the number of primary sampling units (PSUs or clusters) minus the number of strata (see Chapter 10 for more information on PSUs, strata, and sample designs). For replicate-based objects, the degrees of freedom are calculated as one less than the rank of the matrix of replicate weight, where the number of replicates is typically the rank. Note that specifying df = Inf is equivalent to using a normal (z-based) confidence interval – this is the default in {survey}.

These variability types are the same for most of the survey functions, and we provide examples using different variability types throughout this chapter.

5.2.2   Examples

Example 1: Estimated population count

If we want to obtain the estimated number of households in the U.S. (the population of interest) using the Residential Energy Consumption Survey (RECS) data, we can use `survey_count()`. If we do not specify any variables in the `survey_count()` function, it outputs the estimated population count (n) and its corresponding standard error (n_se).

```
recs_des %>%
  survey_count()
```

```
## # A tibble: 1 x 2
##            n  n_se
##        <dbl> <dbl>
## 1 123529025. 0.148
```

Based on this calculation, the estimated number of households in the U.S. is 123,529,025.

Alternatively, we could also use the `survey_tally()` function. The example below yields the same results as `survey_count()`.

```
recs_des %>%
  survey_tally()
```

```
## # A tibble: 1 x 2
##            n  n_se
##        <dbl> <dbl>
## 1 123529025. 0.148
```

Example 2: Estimated counts by subgroups (cross-tabs)

To calculate the estimated number of observations for specific subgroups, such as Region and Division, we can include the variables of interest in the `survey_count()` function. In the example below, we calculate the estimated number of housing units by region and division. The argument `name =` in `survey_count()` allows us to change the name of the count variable in the output from the default n to N.

```
recs_des %>%
  survey_count(Region, Division, name = "N")
```

```
## # A tibble: 10 x 4
##     Region    Division                   N      N_se
##     <fct>     <fct>                   <dbl>     <dbl>
##  1 Northeast New England          5876166  0.0000000137
##  2 Northeast Middle Atlantic     16043503  0.0000000487
##  3 Midwest   East North Central  18546912  0.000000437
##  4 Midwest   West North Central   8495815  0.0000000177
##  5 South     South Atlantic      24843261  0.0000000418
##  6 South     East South Central   7380717. 0.114
##  7 South     West South Central  14619094  0.000488
##  8 West      Mountain North       4615844  0.119
##  9 West      Mountain South       4602070  0.0000000492
## 10 West      Pacific             18505643. 0.00000295
```

When we run the cross-tab, we see that there are an estimated 5,876,166 housing units in the New England Division.

The code results in an error if we try to use the survey_count() syntax with survey_tally():

```
recs_des %>%
  survey_tally(Region, Division, name = "N")
```

```
## Error in `dplyr::summarise()`:
## i In argument: `N = survey_total(Region, vartype = vartype,
##     na.rm = TRUE)`.
## Caused by error:
## ! Factor not allowed in survey functions, should be used as
##     a grouping variable.
```

Use a group_by() function prior to using survey_tally() to successfully run the cross-tab:

```
recs_des %>%
  group_by(Region, Division) %>%
  survey_tally(name = "N")
```

```
## # A tibble: 10 x 4
## # Groups:   Region [4]
##    Region    Division                    N        N_se
##    <fct>     <fct>                   <dbl>       <dbl>
##  1 Northeast New England          5876166  0.0000000137
##  2 Northeast Middle Atlantic     16043503  0.0000000487
##  3 Midwest   East North Central  18546912  0.000000437
##  4 Midwest   West North Central   8495815  0.0000000177
##  5 South     South Atlantic      24843261  0.0000000418
##  6 South     East South Central   7380717. 0.114
##  7 South     West South Central  14619094  0.000488
##  8 West      Mountain North       4615844  0.119
##  9 West      Mountain South       4602070  0.0000000492
## 10 West      Pacific             18505643. 0.00000295
```

5.3   Totals and sums

The `survey_total()` function is analogous to `sum`. It can be applied to continuous variables to obtain the estimated total quantity in a population. Starting from this point in the chapter, all the introduced functions must be called within `summarize()`.

5.3.1   Syntax

Here is the syntax:

```
survey_total(
  x,
  na.rm = FALSE,
  vartype = c("se", "ci", "var", "cv"),
  level = 0.95,
  deff = FALSE,
  df = NULL
)
```

The arguments are:

- x: a variable, expression, or empty
- na.rm: an indicator of whether missing values should be dropped, defaults to FALSE

- vartype: type(s) of variation estimate to calculate including any of c("se", "ci", "var", "cv"), defaults to se (standard error) (see Section 5.2.1 for more information)
- level: a number or a vector indicating the confidence level, defaults to 0.95
- deff: a logical value stating whether the design effect should be returned, defaults to FALSE (this is described in more detail in Section 5.9.3)
- df: (for vartype = 'ci'), a numeric value indicating degrees of freedom for the t-distribution

5.3.2   Examples

Example 1: Estimated population count

To calculate a population count estimate with survey_total(), we leave the argument x empty, as shown in the example below:

```
recs_des %>%
  summarize(Tot = survey_total())
```

```
## # A tibble: 1 x 2
##           Tot Tot_se
##         <dbl>  <dbl>
## 1 123529025.   0.148
```

The estimated number of households in the U.S. is 123,529,025. Note that this result obtained from survey_total() is equivalent to the ones from the survey_count() and survey_tally() functions. However, the survey_total() function is called within summarize(), whereas survey_count() and survey_tally() are not.

Example 2: Overall summation of continuous variables

The distinction between survey_total() and survey_count() becomes more evident when working with continuous variables. Let's compute the total cost of electricity in whole dollars from variable DOLLAREL[1].

```
recs_des %>%
  summarize(elec_bill = survey_total(DOLLAREL))
```

---

[1]RECS has two components: a household survey and an energy supplier survey. For each household that responds, their energy providers are contacted to obtain their energy consumption and expenditure. This value reflects the dollars spent on electricity in 2020, according to the energy supplier. See U.S. Energy Information Administration (2023a) for more details.

```
## # A tibble: 1 x 2
##        elec_bill elec_bill_se
##            <dbl>        <dbl>
## 1 170473527909.    664893504.
```

It is estimated that American residential households spent a total of $170,473,527,909 on electricity in 2020, and the estimate has a standard error of $664,893,504.

Example 3: Summation by groups

Since we are using the {srvyr} package, we can use `group_by()` to calculate the cost of electricity for different groups. Let's examine the variations in the cost of electricity in whole dollars across regions and display the confidence interval instead of the default standard error.

```
recs_des %>%
  group_by(Region) %>%
  summarize(elec_bill = survey_total(DOLLAREL,
    vartype = "ci"
  ))
```

```
## # A tibble: 4 x 4
##     Region        elec_bill elec_bill_low elec_bill_upp
##     <fct>             <dbl>         <dbl>         <dbl>
## 1 Northeast   29430369947.   28788987554.   30071752341.
## 2 Midwest     34972544751.   34339576041.   35605513460.
## 3 South       72496840204.   71534780902.   73458899506.
## 4 West        33573773008.   32909111702.   34238434313.
```

The survey results estimate that households in the Northeast spent $29,430,369,947 with a confidence interval of ($28,788,987,554, $30,071,752,341) on electricity in 2020, while households in the South spent an estimated $72,496,840,204 with a confidence interval of ($71,534,780,902, $73,458,899,506).

As we calculate these numbers, we may notice that the confidence interval of the South is larger than those of other regions. This implies that we have less certainty about the true value of electricity spending in the South. A larger confidence interval could be due to a variety of factors, such as a wider range of electricity spending in the South. We could try to analyze smaller regions within the South to identify areas that are contributing to more variability. Descriptive analyses serve as a valuable starting point for more in-depth exploration and analysis.

5.4 Means and proportions

Means and proportions form the foundation of many research studies. These estimates are often the first things we look for when reviewing research on a given topic. The survey_mean() and survey_prop() functions calculate means and proportions while taking into account the survey design elements. The survey_mean() function should be used on continuous variables of survey data, while the survey_prop() function should be used on categorical variables.

5.4.1 Syntax

The syntax for both means and proportions is very similar:

```
survey_mean(
  x,
  na.rm = FALSE,
  vartype = c("se", "ci", "var", "cv"),
  level = 0.95,
  proportion = FALSE,
  prop_method = c("logit", "likelihood", "asin", "beta", "mean"),
  deff = FALSE,
  df = NULL
)

survey_prop(
  na.rm = FALSE,
  vartype = c("se", "ci", "var", "cv"),
  level = 0.95,
  proportion = TRUE,
  prop_method =
    c("logit", "likelihood", "asin", "beta", "mean", "xlogit"),
  deff = FALSE,
  df = NULL
)
```

Both functions have the following arguments and defaults:

- na.rm: an indicator of whether missing values should be dropped, defaults to FALSE
- vartype: type(s) of variation estimate to calculate including any of c("se", "ci", "var", "cv"), defaults to se (standard error) (see Section 5.2.1 for more information)
- level: a number or a vector indicating the confidence level, defaults to 0.95

- `prop_method`: Method to calculate the confidence interval for confidence intervals
- `deff`: a logical value stating whether the design effect should be returned, defaults to FALSE (this is described in more detail in Section 5.9.3)
- `df`: (for `vartype = 'ci'`), a numeric value indicating degrees of freedom for the t-distribution

There are two main differences in the syntax. The `survey_mean()` function includes the first argument `x`, representing the variable or expression on which the mean should be calculated. The `survey_prop()` does not have an argument to include the variables directly. Instead, prior to `summarize()`, we must use the `group_by()` function to specify the variables of interest for `survey_prop()`. For `survey_mean()`, including a `group_by()` function allows us to obtain the means by different groups.

The other main difference is with the `proportion` argument. The `survey_mean()` function can be used to calculate both means and proportions. Its `proportion` argument defaults to FALSE, indicating it is used for calculating means. If we wish to calculate a proportion using `survey_mean()`, we need to set the `proportion` argument to TRUE. In the `survey_prop()` function, the `proportion` argument defaults to TRUE because the function is specifically designed for calculating proportions.

In Section 5.2.1, we provide an overview of different variability types. The confidence interval used for most measures, such as means and counts, is referred to as a Wald-type interval. However, for proportions, a Wald-type interval with a symmetric t-based confidence interval may not provide accurate coverage, especially when dealing with small sample sizes or proportions "near" 0 or 1. We can use other methods to calculate confidence intervals, which we specify using the `prop_method` option in `survey_prop()`. The options include:

- `logit`: fits a logistic regression model and computes a Wald-type interval on the log-odds scale, which is then transformed to the probability scale. This is the default method.
- `likelihood`: uses the (Rao-Scott) scaled chi-squared distribution for the log-likelihood from a binomial distribution.
- `asin`: uses the variance-stabilizing transformation for the binomial distribution, the arcsine square root, and then back-transforms the interval to the probability scale.
- `beta`: uses the incomplete beta function with an effective sample size based on the estimated variance of the proportion.
- `mean`: the Wald-type interval ($\pm t^*_{df} \times SE$).
- `xlogit`: uses a logit transformation of the proportion, calculates a Wald-type interval, and then back-transforms to the probability scale. This method is the same as those used by default in SUDAAN and SPSS.

Each option yields slightly different confidence interval bounds when dealing with proportions. Please note that when working with `survey_mean()`, we do not need to specify a method unless the `proportion` argument is TRUE. If `proportion` is FALSE, it calculates a symmetric `mean` type of confidence interval.

5.4.2 Examples

Example 1: One variable proportion

If we are interested in obtaining the proportion of people in each region in the RECS data, we can use `group_by()` and `survey_prop()` as shown below:

```
recs_des %>%
  group_by(Region) %>%
  summarize(p = survey_prop())
```

```
## # A tibble: 4 x 3
##   Region    p          p_se
##   <fct>     <dbl>      <dbl>
## 1 Northeast 0.177 0.000000000212
## 2 Midwest   0.219 0.000000000262
## 3 South     0.379 0.000000000740
## 4 West      0.224 0.000000000816
```

17.7% of the households are in the Northeast, 21.9% are in the Midwest, and so on. Note that the proportions in column p add up to one.

The `survey_prop()` function is essentially the same as using `survey_mean()` with a categorical variable and without specifying a numeric variable in the x argument. The following code gives us the same results as above:

```
recs_des %>%
  group_by(Region) %>%
  summarize(p = survey_mean())
```

```
## # A tibble: 4 x 3
##   Region    p          p_se
##   <fct>     <dbl>      <dbl>
## 1 Northeast 0.177 0.000000000212
## 2 Midwest   0.219 0.000000000262
## 3 South     0.379 0.000000000740
## 4 West      0.224 0.000000000816
```

Example 2: Conditional proportions

We can also obtain proportions by more than one variable. In the following
example, we look at the proportion of housing units by Region and whether
air conditioning (A/C) is used (ACUsed)[2].

```
recs_des %>%
  group_by(Region, ACUsed) %>%
  summarize(p = survey_prop())
```

```
## # A tibble: 8 x 4
## # Groups:   Region [4]
##   Region    ACUsed      p    p_se
##   <fct>     <lgl>   <dbl>   <dbl>
## 1 Northeast FALSE  0.110  0.00590
## 2 Northeast TRUE   0.890  0.00590
## 3 Midwest   FALSE  0.0666 0.00508
## 4 Midwest   TRUE   0.933  0.00508
## 5 South     FALSE  0.0581 0.00278
## 6 South     TRUE   0.942  0.00278
## 7 West      FALSE  0.255  0.00759
## 8 West      TRUE   0.745  0.00759
```

When specifying multiple variables, the proportions are conditional. In the
results above, notice that the proportions sum to 1 within each region. This
can be interpreted as the proportion of housing units with A/C within each
region. For example, in the Northeast region, approximately 11.0% of housing
units don't have A/C, while around 89.0% have A/C.

Example 3: Joint proportions

If we're interested in a joint proportion, we use the interact() function. In
the example below, we apply the interact() function to Region and ACUsed:

```
recs_des %>%
  group_by(interact(Region, ACUsed)) %>%
  summarize(p = survey_prop())
```

---

[2]Question text: "Is any air conditioning equipment used in your home?" (U.S. Energy
Information Administration, 2020)

```
## # A tibble: 8 x 4
##    Region     ACUsed      p     p_se
##    <fct>      <lgl>   <dbl>    <dbl>
## 1 Northeast  FALSE  0.0196  0.00105
## 2 Northeast  TRUE   0.158   0.00105
## 3 Midwest    FALSE  0.0146  0.00111
## 4 Midwest    TRUE   0.204   0.00111
## 5 South      FALSE  0.0220  0.00106
## 6 South      TRUE   0.357   0.00106
## 7 West       FALSE  0.0573  0.00170
## 8 West       TRUE   0.167   0.00170
```

In this case, all proportions sum to 1, not just within regions. This means that 15.8% of the population lives in the Northeast and has A/C. As noted earlier, we can use both the survey_prop() and survey_mean() functions, and they produce the same results.

Example 4: Overall mean

Below, we calculate the estimated average cost of electricity in the U.S. using survey_mean(). To include both the standard error and the confidence interval, we can include them in the vartype argument:

```
recs_des %>%
  summarize(elec_bill = survey_mean(DOLLAREL,
    vartype = c("se", "ci")
))
```

```
## # A tibble: 1 x 4
##   elec_bill elec_bill_se elec_bill_low elec_bill_upp
##       <dbl>        <dbl>         <dbl>         <dbl>
## 1     1380.         5.38         1369.         1391.
```

Nationally, the average household spent $1,380 in 2020.

Example 5: Means by subgroup

We can also calculate the estimated average cost of electricity in the U.S. by each region. To do this, we include a group_by() function with the variable of interest before the summarize() function:

```
recs_des %>%
  group_by(Region) %>%
  summarize(elec_bill = survey_mean(DOLLAREL))
```

```
## # A tibble: 4 x 3
##    Region     elec_bill elec_bill_se
##    <fct>          <dbl>        <dbl>
## 1 Northeast      1343.         14.6
## 2 Midwest        1293.         11.7
## 3 South          1548.         10.3
## 4 West           1211.         12.0
```

Households from the West spent approximately \$1,211, while in the South, the average spending was \$1,548.

5.5 Quantiles and medians

To better understand the distribution of a continuous variable like income, we can calculate quantiles at specific points. For example, computing estimates of the quartiles (25%, 50%, 75%) helps us understand how income is spread across the population. We use the `survey_quantile()` function to calculate quantiles in survey data.

Medians are useful for finding the midpoint of a continuous distribution when the data are skewed, as medians are less affected by outliers compared to means. The median is the same as the 50th percentile, meaning the value where 50% of the data are higher and 50% are lower. Because medians are a special, common case of quantiles, we have a dedicated function called `survey_median()` for calculating the median in survey data. Alternatively, we can use the `survey_quantile()` function with the `quantiles` argument set to `0.5` to achieve the same result.

5.5.1 Syntax

The syntax for `survey_quantile()` and `survey_median()` are nearly identical:

```
survey_quantile(
  x,
  quantiles,
  na.rm = FALSE,
  vartype = c("se", "ci", "var", "cv"),
  level = 0.95,
  interval_type =
    c("mean", "beta", "xlogit", "asin", "score", "quantile"),
  qrule = c("math", "school", "shahvaish", "hf1", "hf2", "hf3",
            "hf4", "hf5", "hf6", "hf7", "hf8", "hf9"),
```

```
  df = NULL
)

survey_median(
  x,
  na.rm = FALSE,
  vartype = c("se", "ci", "var", "cv"),
  level = 0.95,
  interval_type =
    c("mean", "beta", "xlogit", "asin", "score", "quantile"),
  qrule = c("math", "school", "shahvaish", "hf1", "hf2", "hf3",
            "hf4", "hf5", "hf6", "hf7", "hf8", "hf9"),
  df = NULL
)
```

The arguments available in both functions are:

- x: a variable, expression, or empty
- na.rm: an indicator of whether missing values should be dropped, defaults to FALSE
- vartype: type(s) of variation estimate to calculate, defaults to se (standard error)
- level: a number or a vector indicating the confidence level, defaults to 0.95
- interval_type: method for calculating a confidence interval
- qrule: rule for defining quantiles. The default is the lower end of the quantile interval ("math"). The midpoint of the quantile interval is the "school" rule. "hf1" to "hf9" are weighted analogs to type=1 to 9 in quantile(). "shahvaish" corresponds to a rule proposed by Shah and Vaish (2006). See vignette("qrule", package="survey") for more information.
- df: (for vartype = 'ci'), a numeric value indicating degrees of freedom for the t-distribution

The only difference between survey_quantile() and survey_median() is the inclusion of the quantiles argument in the survey_quantile() function. This argument takes a vector with values between 0 and 1 to indicate which quantiles to calculate. For example, if we wanted the quartiles of a variable, we would provide quantiles = c(0.25, 0.5, 0.75). While we can specify quantiles of 0 and 1, which represent the minimum and maximum, this is not recommended. It only returns the minimum and maximum of the respondents and cannot be extrapolated to the population, as there is no valid definition of standard error.

In Section 5.2.1, we provide an overview of the different variability types. The interval used in confidence intervals for most measures, such as means and counts, is referred to as a Wald-type interval. However, this is not always the most accurate interval for quantiles. Similar to confidence intervals for proportions, quantiles have various interval types, including asin, beta, mean, and xlogit (see Section 5.4.1). Quantiles also have two more methods available:

- score: the Francisco and Fuller confidence interval based on inverting a score test (only available for design-based survey objects and not replicate-based objects)
- quantile: based on the replicates of the quantile. This is not valid for jackknife-type replicates but is available for bootstrap and BRR replicates.

One note with the score method is that when there are numerous ties in the data, this method may produce confidence intervals that do not contain the estimate. When dealing with a high propensity for ties (e.g., many respondents are the same age), it is recommended to use another method. SUDAAN, for example, uses the score method but adds noise to the values to prevent issues. The documentation in the {survey} package indicates, in general, that the score method may have poorer performance compared to the beta and logit intervals (Lumley, 2010).

5.5.2   Examples

Example 1: Overall quartiles

Quantiles provide insights into the distribution of a variable. Let's look into the quartiles, specifically, the first quartile (p=0.25), the median (p=0.5), and the third quartile (p=0.75) of electric bills.

```
recs_des %>%
  summarize(elec_bill = survey_quantile(DOLLAREL,
    quantiles = c(0.25, .5, 0.75)
  ))
```

```
## # A tibble: 1 x 6
##    elec_bill_q25 elec_bill_q50 elec_bill_q75 elec_bill_q25_se
##            <dbl>         <dbl>         <dbl>            <dbl>
## 1          795.         1215.         1770.             5.69
##    elec_bill_q50_se elec_bill_q75_se
##               <dbl>            <dbl>
## 1              6.33             9.99
```

The output above shows the values for the three quartiles of electric bill costs and their respective standard errors: the 25th percentile is $795 with a standard error of $5.69, the 50th percentile (median) is $1,215 with a standard error of $6.33, and the 75th percentile is $1,770 with a standard error of $9.99.

Example 2: Quartiles by subgroup

We can estimate the quantiles of electric bills by region by using the group_by() function:

```
recs_des %>%
  group_by(Region) %>%
  summarize(elec_bill = survey_quantile(DOLLAREL,
    quantiles = c(0.25, .5, 0.75)
  ))
```

```
## # A tibble: 4 x 7
##   Region    elec_bill_q25 elec_bill_q50 elec_bill_q75 elec_bill_q25_se
##   <fct>             <dbl>         <dbl>         <dbl>            <dbl>
## 1 Northeast          740.         1148.         1712.             13.7
## 2 Midwest            769.         1149.         1632.              8.88
## 3 South              968.         1402.         1945.             10.6
## 4 West               623.         1028.         1568.             10.8
##   elec_bill_q50_se elec_bill_q75_se
##              <dbl>            <dbl>
## 1             16.6             25.8
## 2             11.6             18.6
## 3              9.17            13.9
## 4             14.3             20.5
```

The 25th percentile for the Northeast region is $740, while it is $968 for the South.

Example 3: Minimum and maximum

As mentioned in the syntax section, we can specify quantiles of 0 (minimum) and 1 (maximum), and R calculates these values. However, these are only the minimum and maximum values in the data, and there is not enough information to determine their standard errors:

```
recs_des %>%
  summarize(elec_bill = survey_quantile(DOLLAREL,
    quantiles = c(0, 1)
  ))
```

```
## # A tibble: 1 x 4
##   elec_bill_q00 elec_bill_q100 elec_bill_q00_se elec_bill_q100_se
##           <dbl>          <dbl>            <dbl>             <dbl>
## 1          -889.         15680.              NaN                 0
```

The minimum cost of electricity in the dataset is −$889, while the maximum is $15,680, but the standard error is shown as NaN and 0, respectively. Notice that the minimum cost is a negative number. This may be surprising, but some housing units with solar power sell their energy back to the grid and earn money, which is recorded as a negative expenditure.

Example 4: Overall median

We can calculate the estimated median cost of electricity in the U.S. using the survey_median() function:

```
recs_des %>%
  summarize(elec_bill = survey_median(DOLLAREL))
```

```
## # A tibble: 1 x 2
##   elec_bill elec_bill_se
##       <dbl>        <dbl>
## 1     1215.         6.33
```

Nationally, the median household spent $1,215 in 2020. This is the same result as we obtained using the survey_quantile() function. Interestingly, the average electric bill for households that we calculated in Section 5.4 is $1,380, but the estimated median electric bill is $1,215, indicating the distribution is likely right-skewed.

Example 5: Medians by subgroup

We can calculate the estimated median cost of electricity in the U.S. by region using the group_by() function with the variable(s) of interest before the summarize() function, similar to when we found the mean by region.

```
recs_des %>%
  group_by(Region) %>%
  summarize(elec_bill = survey_median(DOLLAREL))
```

```
## # A tibble: 4 x 3
##   Region    elec_bill elec_bill_se
##   <fct>         <dbl>        <dbl>
## 1 Northeast     1148.        16.6
## 2 Midwest       1149.        11.6
## 3 South         1402.         9.17
## 4 West          1028.        14.3
```

We estimate that households in the Northeast spent a median of $1,148 on electricity, and in the South, they spent a median of $1,402.

5.6 Ratios

A ratio is a measure of the ratio of the sum of two variables, specifically in the form of:

$$\frac{\sum x_i}{\sum y_i}.$$

Note that the ratio is not the same as calculating the following:

$$\frac{1}{N} \sum \frac{x_i}{y_i}$$

which can be calculated with survey_mean() by creating a derived variable $z = x/y$ and then calculating the mean of z.

Say we wanted to assess the energy efficiency of homes in a standardized way, where we can compare homes of different sizes. We can calculate the ratio of energy consumption to the square footage of a home. This helps us meaningfully compare homes of different sizes by identifying how much energy is being used per unit of space. To calculate this ratio, we would run survey_ratio(Energy Consumption in BTUs, Square Footage of Home). If, instead, we used survey_mean(Energy Consumption in BTUs/Square Footage of Home), we would estimate the average energy consumption per square foot of all surveyed homes. While helpful in understanding general energy use, this statistic does not account for differences in home sizes.

5.6.1 Syntax

The syntax for survey_ratio() is as follows:

```
survey_ratio(
  numerator,
  denominator,
  na.rm = FALSE,
  vartype = c("se", "ci", "var", "cv"),
  level = 0.95,
  deff = FALSE,
  df = NULL
)
```

The arguments are:

- numerator: The numerator of the ratio
- denominator: The denominator of the ratio

- `na.rm`: A logical value to indicate whether missing values should be dropped
- `vartype`: type(s) of variation estimate to calculate including any of `c("se",  "ci", "var", "cv")`, defaults to `se` (standard error) (see Section 5.2.1 for more information)
- `level`: A single number or vector of numbers indicating the confidence level
- `deff`: A logical value to indicate whether the design effect should be returned (this is described in more detail in Section 5.9.3)
- `df`: (For vartype = "ci" only) A numeric value indicating the degrees of freedom for t-distribution

5.6.2   Examples

Example 1: Overall ratios

Suppose we wanted to find the ratio of dollars spent on liquid propane per unit (in British thermal unit [Btu]) nationally[3]. To find the average cost to a household, we can use `survey_mean()`. However, to find the national unit rate, we can use `survey_ratio()`. In the following example, we show both methods and discuss the interpretation of each:

```
recs_des %>%
  summarize(
    DOLLARLP_Tot = survey_total(DOLLARLP, vartype = NULL),
    BTULP_Tot = survey_total(BTULP, vartype = NULL),
    DOL_BTU_Rat = survey_ratio(DOLLARLP, BTULP),
    DOL_BTU_Avg = survey_mean(DOLLARLP / BTULP, na.rm = TRUE)
  )
```

```
## # A tibble: 1 x 6
##    DOLLARLP_Tot     BTULP_Tot DOL_BTU_Rat DOL_BTU_Rat_se DOL_BTU_Avg
##           <dbl>         <dbl>       <dbl>          <dbl>       <dbl>
## 1  8122911173. 391425311586.      0.0208       0.000240      0.0240
##    DOL_BTU_Avg_se
##             <dbl>
## 1        0.000223
```

The ratio of the total spent on liquid propane to the total consumption was 0.0208, but the average rate was 0.024. With a bit of calculation, we can show that the ratio is the ratio of the totals `DOLLARLP_Tot/BTULP_Tot`=8,122,911,173/391,425,311,586=0.0208. Although the estimated ratio can be calculated manually in this manner, the standard

---

[3]The value of `DOLLARLP` reflects the annualized amount spent on liquid propane and `BTULP` reflects the annualized consumption in Btu of liquid propane (U.S. Energy Information Administration, 2020).

error requires the use of the `survey_ratio()` function. The average can be interpreted as the average rate paid by a household.

Example 2: Ratios by subgroup

As previously done with other estimates, we can use `group_by()` to examine whether this ratio varies by region.

```
recs_des %>%
  group_by(Region) %>%
  summarize(DOL_BTU_Rat = survey_ratio(DOLLARLP, BTULP)) %>%
  arrange(DOL_BTU_Rat)
```

```
## # A tibble: 4 x 3
##   Region    DOL_BTU_Rat DOL_BTU_Rat_se
##   <fct>           <dbl>          <dbl>
## 1 Midwest        0.0158       0.000240
## 2 South          0.0245       0.000388
## 3 West           0.0246       0.000875
## 4 Northeast      0.0247       0.000488
```

Although not a formal statistical test, it appears that the cost ratios for liquid propane are the lowest in the Midwest (0.0158).

5.7 Correlations

The correlation is a measure of the linear relationship between two continuous variables, which ranges between –1 and 1. The most commonly used method is Pearson's correlation (referred to as correlation henceforth). A sample correlation for a simple random sample is calculated as follows:

$$\frac{\sum(x_i - \bar{x})(y_i - \bar{y})}{\sqrt{\sum(x_i - \bar{x})^2}\sqrt{\sum(y_i - \bar{y})^2}}$$

When using `survey_corr()` for designs other than a simple random sample, the weights are applied when estimating the correlation.

5.7.1 Syntax

The syntax for `survey_corr()` is as follows:

```
survey_corr(
  x,
  y,
  na.rm = FALSE,
  vartype = c("se", "ci", "var", "cv"),
  level = 0.95,
  df = NULL
)
```

The arguments are:

- `x`: A variable or expression
- `y`: A variable or expression
- `na.rm`: A logical value to indicate whether missing values should be dropped
- `vartype`: Type(s) of variation estimate to calculate including any of `c("se", "ci", "var", "cv")`, defaults to `se` (standard error) (see Section 5.2.1 for more information)
- `level`: (For vartype = "ci" only) A single number or vector of numbers indicating the confidence level
- `df`: (For vartype = "ci" only) A numeric value indicating the degrees of freedom for t-distribution

5.7.2 Examples

Example 1: Overall correlation

We can calculate the correlation between the total square footage of homes (`TOTSQFT_EN`)[4] and electricity consumption (`BTUEL`)[5].

```
recs_des %>%
  summarize(SQFT_Elec_Corr = survey_corr(TOTSQFT_EN, BTUEL))
```

```
## # A tibble: 1 x 2
##   SQFT_Elec_Corr SQFT_Elec_Corr_se
##            <dbl>             <dbl>
## 1          0.417           0.00689
```

---

[4]Question text: "What is the square footage of your home?" (U.S. Energy Information Administration, 2020)

[5]BTUEL is derived from the supplier side component of the survey where BTUEL represents the electricity consumption in British thermal units (Btus) converted from kilowatt hours (kWh) in a year (U.S. Energy Information Administration, 2020).

The correlation between the total square footage of homes and electricity consumption is 0.417, indicating a moderate positive relationship.

Example 2: Correlations by subgroup

We can explore the correlation between total square footage and electricity consumption based on subgroups, such as whether A/C is used (ACUsed).

```
recs_des %>%
  group_by(ACUsed) %>%
  summarize(SQFT_Elec_Corr = survey_corr(TOTSQFT_EN, DOLLAREL))
```

```
## # A tibble: 2 x 3
##   ACUsed SQFT_Elec_Corr SQFT_Elec_Corr_se
##   <lgl>           <dbl>             <dbl>
## 1 FALSE           0.290           0.0240
## 2 TRUE            0.401           0.00808
```

For homes without A/C, there is a small positive correlation between total square footage with electricity consumption (0.29). For homes with A/C, the correlation of 0.401 indicates a stronger positive correlation between total square footage and electricity consumption.

5.8 Standard deviation and variance

All survey functions produce an estimate of the variability of a given estimate. No additional function is needed when dealing with variable estimates. However, if we are specifically interested in population variance and standard deviation, we can use the survey_var() and survey_sd() functions. In our experience, it is not common practice to use these functions. They can be used when designing a future study to gauge population variability and inform sampling precision.

5.8.1 Syntax

As with non-survey data, the standard deviation estimate is the square root of the variance estimate. Therefore, the survey_var() and survey_sd() functions share the same arguments, except the standard deviation does not allow the usage of vartype.

```
survey_var(
  x,
  na.rm = FALSE,
  vartype = c("se", "ci", "var"),
  level = 0.95,
  df = NULL
)

survey_sd(
  x,
  na.rm = FALSE
)
```

The arguments are:

- x: A variable or expression, or empty
- na.rm: A logical value to indicate whether missing values should be dropped
- vartype: Type(s) of variation estimate to calculate including any of c("se", "ci", "var"), defaults to se (standard error) (see Section 5.2.1 for more information)
- level: (For vartype = "ci" only) A single number or vector of numbers indicating the confidence level
- df: (For vartype = "ci" only) A numeric value indicating the degrees of freedom for t-distribution

5.8.2   Examples

Example 1: Overall variability

Let's return to electricity bills and explore the variability in electricity expenditure.

```
recs_des %>%
  summarize(
    var_elbill = survey_var(DOLLAREL),
    sd_elbill = survey_sd(DOLLAREL)
  )
```

```
## # A tibble: 1 x 3
##   var_elbill var_elbill_se sd_elbill
##        <dbl>         <dbl>     <dbl>
## 1    704906.        13926.      840.
```

We may encounter a warning related to deprecated underlying calculations performed by the survey_var() function. This warning is a result of changes in the way R handles recycling in vectorized operations. The results are still valid. They give an estimate of the population variance of electricity bills (`var_elbill`), the standard error of that variance (`var_elbill_se`), and the estimated population standard deviation of electricity bills (`sd_elbill`). Note that no standard error is associated with the standard deviation; this is the only estimate that does not include a standard error.

Example 2: Variability by subgroup

To find out if the variability in electricity expenditure is similar across regions, we can calculate the variance by region using `group_by()`:

```
recs_des %>%
  group_by(Region) %>%
  summarize(
    var_elbill = survey_var(DOLLAREL),
    sd_elbill = survey_sd(DOLLAREL)
  )
```

```
## # A tibble: 4 x 4
##   Region    var_elbill var_elbill_se sd_elbill
##   <fct>          <dbl>         <dbl>     <dbl>
## 1 Northeast    775450.        38843.      881.
## 2 Midwest      552423.        25252.      743.
## 3 South        702521.        30641.      838.
## 4 West         717886.        30597.      847.
```

5.9 Additional topics

5.9.1 Unweighted analysis

Sometimes, it is helpful to calculate an unweighted estimate of a given variable. For this, we use the `unweighted()` function in the `summarize()` function. The `unweighted()` function calculates unweighted summaries from a `tbl_svy` object, providing the summary among the respondents without extrapolating to a population estimate. The `unweighted()` function can be used in conjunction with any {dplyr} functions. Here is an example looking at the average household electricity cost:

```
recs_des %>%
  summarize(
    elec_bill = survey_mean(DOLLAREL),
```

```
    elec_unweight = unweighted(mean(DOLLAREL))
  )
```

```
## # A tibble: 1 x 3
##   elec_bill elec_bill_se elec_unweight
##       <dbl>        <dbl>         <dbl>
## 1     1380.         5.38         1425.
```

It is estimated that American residential households spent an average of $1,380 on electricity in 2020, and the estimate has a standard error of $5.38. The `unweighted()` function calculates the unweighted average and represents the average amount of money spent on electricity in 2020 by the respondents, which was $1,425.

5.9.2 Subpopulation analysis

We mentioned using `filter()` to subset a survey object for analysis. This operation should be done after creating the survey design object. Subsetting data before creating the object can lead to incorrect variability estimates, if subsetting removes an entire Primary Sampling Unit (PSU; see Chapter 10 for more information on PSUs and sample designs).

Suppose we want estimates of the average amount spent on natural gas among housing units using natural gas (based on the variable BTUNG)[6]. We first filter records to only include records where BTUNG > 0 and then find the average amount spent.

```
recs_des %>%
  filter(BTUNG > 0) %>%
  summarize(NG_mean = survey_mean(DOLLARNG,
    vartype = c("se", "ci")
))
```

```
## # A tibble: 1 x 4
##   NG_mean NG_mean_se NG_mean_low NG_mean_upp
##     <dbl>      <dbl>       <dbl>       <dbl>
## 1    631.       4.64        621.        640.
```

The estimated average amount spent on natural gas among households that use natural gas is $631. Let's compare this to the mean when we do not filter.

---

[6]BTUNG is derived from the supplier side component of the survey where BTUNG represents the natural gas consumption in British thermal units (Btus) in a year (U.S. Energy Information Administration, 2020).

```
recs_des %>%
  summarize(NG_mean = survey_mean(DOLLARNG,
    vartype = c("se", "ci")
  ))
```

```
## # A tibble: 1 x 4
##   NG_mean NG_mean_se NG_mean_low NG_mean_upp
##     <dbl>      <dbl>       <dbl>       <dbl>
## 1    382.       3.41        375.        389.
```

Based on this calculation, the estimated average amount spent on natural gas is \$382. Note that applying the filter to include only housing units that use natural gas yields a higher mean than when not applying the filter. This is because including housing units that do not use natural gas introduces many \$0 amounts, impacting the mean calculation.

5.9.3 Design effects

The design effect measures how the precision of an estimate is influenced by the sampling design. In other words, it measures how much more or less statistically efficient the survey design is compared to a simple random sample (SRS). It is computed by taking the ratio of the estimate's variance under the design at hand to the estimate's variance under a simple random sample without replacement. A design effect less than 1 indicates that the design is more statistically efficient than an SRS design, which is rare but possible in a stratified sampling design where the outcome correlates with the stratification variable(s). A design effect greater than 1 indicates that the design is less statistically efficient than an SRS design. From a design effect, we can calculate the effective sample size as follows:

$$n_{eff} = \frac{n}{D_{eff}}$$

where n is the nominal sample size (the number of survey responses) and D_{eff} is the estimated design effect. We can interpret the effective sample size n_{eff} as the hypothetical sample size that a survey using an SRS design would need to achieve the same precision as the design at hand. Design effects specific to each outcome — outcomes that are less clustered in the population have smaller design effects than outcomes that are clustered.

In the {srvyr} package, design effects can be calculated for totals, proportions, means, and ratio estimates by setting the deff argument to TRUE in the corresponding functions. In the example below, we calculate the design effects for the average consumption of electricity (BTUEL), natural gas (BTUNG), liquid propane (BTULP), fuel oil (BTUFO), and wood (BTUWOOD) by setting deff = TRUE:

```
recs_des %>%
  summarize(across(
    c(BTUEL, BTUNG, BTULP, BTUFO, BTUWOOD),
    ~ survey_mean(.x, deff = TRUE, vartype = NULL)
  )) %>%
  select(ends_with("deff"))
```

```
## # A tibble: 1 x 5
##   BTUEL_deff BTUNG_deff BTULP_deff BTUFO_deff BTUWOOD_deff
##        <dbl>      <dbl>      <dbl>      <dbl>        <dbl>
## 1      0.597      0.938       1.21      0.720         1.10
```

For the values less than 1 (BTUEL_deff and BTUFO_deff), the results suggest
that the survey design is more efficient than a simple random sample. For the
values greater than 1 (BTUNG_deff, BTULP_deff, and BTUWOOD_deff), the results
indicate that the survey design is less efficient than a simple random sample.

5.9.4   Creating summary rows

When using group_by() in analysis, the results are returned with a row for
each group or combination of groups. Often, we want both breakdowns by
group and a summary row for the estimate representing the entire population.
For example, we may want the average electricity consumption by region and
nationally. The {srvyr} package has the convenient cascade() function, which
adds summary rows for the total of a group. It is used instead of summarize()
and has similar functionalities along with some additional features.

Syntax

The syntax is as follows:

```
cascade(
  .data,
  ...,
  .fill = NA,
  .fill_level_top = FALSE,
  .groupings = NULL
)
```

where the arguments are:

- .data: A tbl_svy object
- ...: Name-value pairs of summary functions (same as the summarize() function)
- .fill: Value to fill in for group summaries (defaults to NA)

- .fill_level_top: When filling factor variables, whether to put the value '.fill' in the first position (defaults to FALSE, placing it in the bottom)

Example

First, let's look at an example where we calculate the average household electricity cost. Then, we build on it to examine the features of the cascade() function. In the first example below, we calculate the average household energy cost DOLLAREL_mn using survey_mean() without modifying any of the argument defaults in the function:

```
recs_des %>%
  cascade(DOLLAREL_mn = survey_mean(DOLLAREL))
```

```
## # A tibble: 1 x 2
##   DOLLAREL_mn DOLLAREL_mn_se
##         <dbl>          <dbl>
## 1       1380.           5.38
```

Next, let's group the results by region by adding group_by() before the cascade() function:

```
recs_des %>%
  group_by(Region) %>%
  cascade(DOLLAREL_mn = survey_mean(DOLLAREL))
```

```
## # A tibble: 5 x 3
##   Region    DOLLAREL_mn DOLLAREL_mn_se
##   <fct>           <dbl>          <dbl>
## 1 Northeast       1343.           14.6
## 2 Midwest         1293.           11.7
## 3 South           1548.           10.3
## 4 West            1211.           12.0
## 5 <NA>            1380.           5.38
```

We can see the estimated average electricity bills by region: $1,343 for the Northeast, $1,548 for the South, and so on. The last row, where Region = NA, is the national average electricity bill, $1,380. However, naming the national "region" as NA is not very informative. We can give it a better name using the .fill argument.

```
recs_des %>%
  group_by(Region) %>%
  cascade(
    DOLLAREL_mn = survey_mean(DOLLAREL),
```

```
   .fill = "National"
  )
```

```
## # A tibble: 5 x 3
##   Region     DOLLAREL_mn DOLLAREL_mn_se
##   <fct>          <dbl>        <dbl>
## 1 Northeast      1343.         14.6
## 2 Midwest        1293.         11.7
## 3 South          1548.         10.3
## 4 West           1211.         12.0
## 5 National       1380.          5.38
```

We can move the summary row to the first row by adding `.fill_level_top =
TRUE` to `cascade()`:

```
recs_des %>%
  group_by(Region) %>%
  cascade(
    DOLLAREL_mn = survey_mean(DOLLAREL),
    .fill = "National",
    .fill_level_top = TRUE
  )
```

```
## # A tibble: 5 x 3
##   Region     DOLLAREL_mn DOLLAREL_mn_se
##   <fct>          <dbl>        <dbl>
## 1 National       1380.          5.38
## 2 Northeast      1343.         14.6
## 3 Midwest        1293.         11.7
## 4 South          1548.         10.3
## 5 West           1211.         12.0
```

While the results remain the same, the table is now easier to interpret.

5.9.5   Calculating estimates for many outcomes

Often, we are interested in a summary statistic across many variables. Useful
tools include the `across()` function in {dplyr}, shown a few times above, and
the `map()` function in {purrr}.

The `across()` function applies the same function to multiple columns within
`summarize()`. This works well with all functions shown above, except for
`survey_prop()`. In a later example, we tackle summarizing multiple proportions.

Example 1: `across()`

Suppose we want to calculate the total and average consumption, along with coefficients of variation (CV), for each fuel type. These include the reported consumption of electricity (BTUEL), natural gas (BTUNG), liquid propane (BTULP), fuel oil (BTUFO), and wood (BTUWOOD), as mentioned in the section on design effects. We can take advantage of the fact that these are the only variables that start with "BTU" by selecting them with `starts_with("BTU")` in the `across()` function. For each selected column (`.x`), `across()` creates a list of two functions to be applied: `survey_total()` to calculate the total and `survey_mean()` to calculate the mean, along with their CV (`vartype = "cv"`). Finally, `.unpack = "{outer}.{inner}"` specifies that the resulting column names are a concatenation of the variable name, followed by Total or Mean, and then "coef" or "cv."

```
consumption_ests <- recs_des %>%
  summarize(across(
    starts_with("BTU"),
    list(
      Total = ~ survey_total(.x, vartype = "cv"),
      Mean = ~ survey_mean(.x, vartype = "cv")
    ),
    .unpack = "{outer}.{inner}"
  ))

consumption_ests
```

```
## # A tibble: 1 x 20
##    BTUEL_Total.coef BTUEL_Total._cv BTUEL_Mean.coef BTUEL_Mean._cv
##               <dbl>           <dbl>           <dbl>          <dbl>
## 1      4453284510065         0.00377          36051.        0.00377
## # i 16 more variables: BTUNG_Total.coef <dbl>, BTUNG_Total._cv <dbl>,
## #   BTUNG_Mean.coef <dbl>, BTUNG_Mean._cv <dbl>,
## #   BTULP_Total.coef <dbl>, BTULP_Total._cv <dbl>,
## #   BTULP_Mean.coef <dbl>, BTULP_Mean._cv <dbl>,
## #   BTUFO_Total.coef <dbl>, BTUFO_Total._cv <dbl>,
## #   BTUFO_Mean.coef <dbl>, BTUFO_Mean._cv <dbl>,
## #   BTUWOOD_Total.coef <dbl>, BTUWOOD_Total._cv <dbl>, ...
```

The estimated total consumption of electricity (BTUEL) is 4,453,284,510,065 (`BTUEL_Total.coef`), the estimated average consumption is 36,051 (`BTUEL_Mean.coef`), and the CV is 0.0038.

In the example above, the table was quite wide. We may prefer a row for each fuel type. Using the `pivot_longer()` and `pivot_wider()` functions from {tidyr} can help us achieve this. First, we use `pivot_longer()` to make each variable a column, changing the data to a "long" format. We use the `names_to`

argument to specify new column names: FuelType, Stat, and Type. Then, the
names_pattern argument extracts the names in the original column names
based on the regular expression pattern BTU(.*)_(.*)\\.(.*). They are saved
in the column names defined in names_to.

```
consumption_ests_long <- consumption_ests %>%
  pivot_longer(
    cols = everything(),
    names_to = c("FuelType", "Stat", "Type"),
    names_pattern = "BTU(.*)_(.*)\\.(.*)"
  )

consumption_ests_long
```

```
## # A tibble: 20 x 4
##    FuelType Stat  Type           value
##    <chr>    <chr> <chr>          <dbl>
##  1 EL       Total coef  4453284510065
##  2 EL       Total _cv        0.00377
##  3 EL       Mean  coef         36051.
##  4 EL       Mean  _cv         0.00377
##  5 NG       Total coef  4240769382106.
##  6 NG       Total _cv        0.00908
##  7 NG       Mean  coef         34330.
##  8 NG       Mean  _cv         0.00908
##  9 LP       Total coef   391425311586.
## 10 LP       Total _cv         0.0380
## 11 LP       Mean  coef          3169.
## 12 LP       Mean  _cv          0.0380
## 13 FO       Total coef   395699976655.
## 14 FO       Total _cv         0.0343
## 15 FO       Mean  coef          3203.
## 16 FO       Mean  _cv          0.0343
## 17 WOOD     Total coef   345091088404.
## 18 WOOD     Total _cv         0.0454
## 19 WOOD     Mean  coef          2794.
## 20 WOOD     Mean  _cv          0.0454
```

Then, we use pivot_wider() to create a table that is nearly ready for publication.
Within the function, we can make the names for each element more descriptive
and informative by gluing the Stat and Type together with names_glue. Further
details on creating publication-ready tables are covered in Chapter 8.

```
consumption_ests_long %>%
  mutate(Type = case_when(
    Type == "coef" ~ "",
    Type == "_cv" ~ " (CV)"
  )) %>%
  pivot_wider(
    id_cols = FuelType,
    names_from = c(Stat, Type),
    names_glue = "{Stat}{Type}",
    values_from = value
  )
```

```
## # A tibble: 5 x 5
##   FuelType        Total `Total (CV)`  Mean `Mean (CV)`
##   <chr>           <dbl>        <dbl> <dbl>       <dbl>
## 1 EL      4453284510065      0.00377 36051.     0.00377
## 2 NG     4240769382106.      0.00908 34330.     0.00908
## 3 LP      391425311586.       0.0380  3169.      0.0380
## 4 FO      395699976655.       0.0343  3203.      0.0343
## 5 WOOD    345091088404.       0.0454  2794.      0.0454
```

Example 2: Proportions with `across()`

As mentioned earlier, proportions do not work as well directly with the `across()` method. If we want the proportion of houses with A/C and the proportion of houses with heating, we require two separate `group_by()` statements as shown below:

```
recs_des %>%
  group_by(ACUsed) %>%
  summarize(p = survey_prop())
```

```
## # A tibble: 2 x 3
##   ACUsed     p   p_se
##   <lgl>  <dbl>  <dbl>
## 1 FALSE  0.113 0.00306
## 2 TRUE   0.887 0.00306
```

```
recs_des %>%
  group_by(SpaceHeatingUsed) %>%
  summarize(p = survey_prop())
```

```
## # A tibble: 2 x 3
##   SpaceHeatingUsed        p    p_se
##   <lgl>               <dbl>   <dbl>
## 1 FALSE              0.0469 0.00207
## 2 TRUE               0.953  0.00207
```

We estimate 88.7% of households have A/C and 95.3% have heating.

If we are only interested in the TRUE outcomes, that is, the proportion of households that have A/C and the proportion that have heating, we can simplify the code. Applying survey_mean() to a logical variable is the same as using survey_prop(), as shown below:

```
cool_heat_tab <- recs_des %>%
  summarize(across(c(ACUsed, SpaceHeatingUsed), ~ survey_mean(.x),
    .unpack = "{outer}.{inner}"
  ))

cool_heat_tab
```

```
## # A tibble: 1 x 4
##   ACUsed.coef ACUsed._se SpaceHeatingUsed.coef SpaceHeatingUsed._se
##         <dbl>      <dbl>                 <dbl>                <dbl>
## 1       0.887    0.00306                 0.953              0.00207
```

Note that the estimates are the same as those obtained using the separate group_by() statements. As before, we can use pivot_longer() to structure the table in a more suitable format for distribution.

```
cool_heat_tab %>%
  pivot_longer(everything(),
    names_to = c("Comfort", ".value"),
    names_pattern = "(.*)\\.(.*)"
  ) %>%
  rename(
    p = coef,
    se = `_se`
  )
```

```
## # A tibble: 2 x 3
##   Comfort              p      se
##   <chr>            <dbl>   <dbl>
## 1 ACUsed           0.887 0.00306
## 2 SpaceHeatingUsed 0.953 0.00207
```

Example 3: `purrr::map()`

Loops are a common tool when dealing with repetitive calculations. The {purrr} package provides the `map()` functions, which, like a loop, allow us to perform the same task across different elements (Wickham and Henry, 2023). In our case, we may want to calculate proportions from the same design multiple times. A straightforward approach is to design the calculation for one variable, build a function based on that, and then apply it iteratively for the rest of the variables.

Suppose we want to create a table that shows the proportion of people who express trust in their government (`TrustGovernment`)[7] as well as those that trust in people (`TrustPeople`)[8] using data from the 2020 ANES.

First, we create a table for a single variable. The table includes the variable name as a column, the response, and the corresponding percentage with its standard error.

```
anes_des %>%
  drop_na(TrustGovernment) %>%
  group_by(TrustGovernment) %>%
  summarize(p = survey_prop() * 100) %>%
  mutate(Variable = "TrustGovernment") %>%
  rename(Answer = TrustGovernment) %>%
  select(Variable, everything())
```

```
## # A tibble: 5 x 4
##     Variable        Answer                p  p_se
##     <chr>           <fct>             <dbl> <dbl>
## 1 TrustGovernment Always             1.55 0.204
## 2 TrustGovernment Most of the time  13.2  0.553
## 3 TrustGovernment About half the time 30.9 0.829
## 4 TrustGovernment Some of the time  43.4  0.855
## 5 TrustGovernment Never             11.0  0.566
```

We estimate that 1.55% of people always trust the government, 13.16% trust the government most of the time, and so on.

Now, we want to use the original series of steps as a template to create a general function `calcps()` that can apply the same steps to other variables. We replace `TrustGovernment` with an argument for a generic variable, var. Referring to

---

[7]Question text: "How often can you trust the federal government in Washington to do what is right? (Always, most of the time, about half the time, some of the time, or never)" (American National Election Studies, 2021)

[8]Question text: "Generally speaking, how often can you trust other people? (Always, most of the time, about half the time, some of the time, or never)" (American National Election Studies, 2021)

var involves a bit of tidy evaluation, an advanced skill. To learn more, we recommend Wickham (2019).

```r
calcps <- function(var) {
  anes_des %>%
    drop_na(!!sym(var)) %>%
    group_by(!!sym(var)) %>%
    summarize(p = survey_prop() * 100) %>%
    mutate(Variable = var) %>%
    rename(Answer := !!sym(var)) %>%
    select(Variable, everything())
}
```

We then apply this function to the two variables of interest, TrustGovernment and TrustPeople:

```r
calcps("TrustGovernment")
```

```
## # A tibble: 5 x 4
##   Variable        Answer                 p  p_se
##   <chr>           <fct>              <dbl> <dbl>
## 1 TrustGovernment Always              1.55 0.204
## 2 TrustGovernment Most of the time   13.2  0.553
## 3 TrustGovernment About half the time 30.9 0.829
## 4 TrustGovernment Some of the time   43.4  0.855
## 5 TrustGovernment Never              11.0  0.566
```

```r
calcps("TrustPeople")
```

```
## # A tibble: 5 x 4
##   Variable    Answer                  p  p_se
##   <chr>       <fct>               <dbl> <dbl>
## 1 TrustPeople Always              0.809 0.164
## 2 TrustPeople Most of the time   41.4   0.857
## 3 TrustPeople About half the time 28.2  0.776
## 4 TrustPeople Some of the time   24.5   0.670
## 5 TrustPeople Never               5.05  0.422
```

Finally, we use map() to iterate over as many variables as needed. We feed our desired variables into map() along with our custom function, calcps. The output is a tibble with the variable names in the "Variable" column, the responses in the "Answer" column, along with the percentage and standard error. The list_rbind() function combines the rows into a single tibble. This example extends nicely when dealing with numerous variables for which we want percentage estimates.

```
c("TrustGovernment", "TrustPeople") %>%
  map(calcps) %>%
  list_rbind()
```

```
## # A tibble: 10 x 4
##    Variable        Answer                   p  p_se
##    <chr>           <fct>                <dbl> <dbl>
## 1  TrustGovernment Always                1.55 0.204
## 2  TrustGovernment Most of the time      13.2 0.553
## 3  TrustGovernment About half the time   30.9 0.829
## 4  TrustGovernment Some of the time      43.4 0.855
## 5  TrustGovernment Never                 11.0 0.566
## 6  TrustPeople     Always               0.809 0.164
## 7  TrustPeople     Most of the time      41.4 0.857
## 8  TrustPeople     About half the time   28.2 0.776
## 9  TrustPeople     Some of the time      24.5 0.670
## 10 TrustPeople     Never                 5.05 0.422
```

In addition to our results above, we can also see the output for `TrustPeople`. While we estimate that 1.55% of people always trust the government, 0.81% always trust people.

5.10 Exercises

The exercises use the design objects `anes_des` and `recs_des` provided in the Prerequisites box at the beginning of the chapter.

1. How many females have a graduate degree? Hint: The variables `Gender` and `Education` will be useful.

2. What percentage of people identify as "Strong Democrat"? Hint: The variable `PartyID` indicates someone's party affiliation.

3. What percentage of people who voted in the 2020 election identify as "Strong Republican"? Hint: The variable `VotedPres2020` indicates whether someone voted in 2020.

4. What percentage of people voted in both the 2016 election and the 2020 election? Include the logit confidence interval. Hint: The variable `VotedPres2016` indicates whether someone voted in 2016.

5. What is the design effect for the proportion of people who voted early? Hint: The variable `EarlyVote2020` indicates whether someone voted early in 2020.

6. What is the median temperature people set their thermostats to at night during the winter? Hint: The variable `WinterTempNight` indicates the temperature that people set their thermostat to in the winter at night.

7. People sometimes set their temperature differently over different seasons and during the day. What median temperatures do people set their thermostats to in the summer and winter, both during the day and at night? Include confidence intervals. Hint: Use the variables `WinterTempDay`, `WinterTempNight`, `SummerTempDay`, and `SummerTempNight`.

8. What is the correlation between the temperature that people set their temperature at during the night and during the day in the summer?

9. What is the 1st, 2nd, and 3rd quartile of money spent on energy by Building America (BA) climate zone? Hint: `TOTALDOL` indicates the total amount spent on all fuel, and `ClimateRegion_BA` indicates the BA climate zones.

6

Statistical testing

Prerequisites

For this chapter, load the following packages:

```r
library(tidyverse)
library(survey)
library(srvyr)
library(srvyrexploR)
library(broom)
library(gt)
library(prettyunits)
```

We are using data from ANES and RECS described in Chapter 4. As a reminder, here is the code to create the design objects for each to use throughout this chapter. For ANES, we need to adjust the weight so it sums to the population instead of the sample (see the ANES documentation and Chapter 4 for more information).

```r
targetpop <- 231592693

anes_adjwgt <- anes_2020 %>%
  mutate(Weight = Weight / sum(Weight) * targetpop)

anes_des <- anes_adjwgt %>%
  as_survey_design(
    weights = Weight,
    strata = Stratum,
    ids = VarUnit,
    nest = TRUE
  )
```

For RECS, details are included in the RECS documentation and Chapters 4 and 10.

```
recs_des <- recs_2020 %>%
  as_survey_rep(
    weights = NWEIGHT,
    repweights = NWEIGHT1:NWEIGHT60,
    type = "JK1",
    scale = 59 / 60,
    mse = TRUE
  )
```

6.1 Introduction

When analyzing survey results, the point estimates described in Chapter 5 help us understand the data at a high level. Still, we often want to make comparisons between different groups. These comparisons are calculated through statistical testing.

The general idea of statistical testing is the same for data obtained through surveys and data obtained through other methods, where we compare the point estimates and uncertainty estimates of each statistic to see if statistically significant differences exist. However, statistical testing for complex surveys involves additional considerations due to the need to account for the sampling design in order to obtain accurate uncertainty estimates.

Statistical testing, also called hypothesis testing, involves declaring a null and alternative hypothesis. A null hypothesis is denoted as H_0 and the alternative hypothesis is denoted as H_A. The null hypothesis is the default assumption in that there are no differences in the data, or that the data are operating under "standard" behaviors. On the other hand, the alternative hypothesis is the break from the "standard," and we are trying to determine if the data support this alternative hypothesis.

Let's review an example outside of survey data. If we are flipping a coin, a null hypothesis would be that the coin is fair and that each side has an equal chance of being flipped. In other words, the probability of the coin landing on each side is $1/2$, whereas an alternative hypothesis could be that the coin is unfair and that one side has a higher probability of being flipped (e.g., a probability of $1/4$ to get heads but a probability of $3/4$ to get tails). We write this set of hypotheses as:

- $H_0 : \rho_{heads} = \rho_{tails}$, where ρ_x is the probability of flipping the coin and having it land on heads (ρ_{heads}) or tails (ρ_{tails})
- $H_A : \rho_{heads} \neq \rho_{tails}$

When we conduct hypothesis testing, the statistical models calculate a p-value, which shows how likely we are to observe the data if the null hypothesis is true. If the p-value (a probability between 0 and 1) is small, we have strong evidence to reject the null hypothesis, as it is unlikely to see the data we observe if the null hypothesis is true. However, if the p-value is large, we say we do not have evidence to reject the null hypothesis. The size of the p-value for this cut-off is determined by Type 1 error known as α. A common Type 1 error value for statistical testing is to use $\alpha = 0.05$[1]. Explanations of statistical testing often refer to confidence level. The confidence level is the inverse of the Type 1 error. Thus, if $\alpha = 0.05$, the confidence level would be 95%.

The functions in the {survey} package allow for the correct estimation of the uncertainty estimates (e.g., standard deviations and confidence intervals). This chapter covers the following statistical tests with survey data and the following functions from the {survey} package (Lumley, 2010):

- Comparison of proportions (svyttest())
- Comparison of means (svyttest())
- Goodness-of-fit tests (svygofchisq())
- Tests of independence (svychisq())
- Tests of homogeneity (svychisq())

6.2 Dot notation

Up to this point, we have shown functions that use wrappers from the {srvyr} package. This means that the functions work with tidyverse syntax. However, the functions in this chapter do not have wrappers in the {srvyr} package and are instead used directly from the {survey} package. Therefore, the design object is not the first argument, and to use these functions with the magrittr pipe (%>%) and tidyverse syntax, we need to use dot (.) notation[2].

Functions that work with the magrittr pipe (%>%) have the dataset as the first argument. When we run a function with the pipe, it automatically places anything to the left of the pipe into the first argument of the function to the right of the pipe. For example, if we wanted to take the towny data from the {gt} package and filter to municipalities with the Census Subdivision Type of "city," we can write the code in at least four different ways:

---

[1] For more information on statistical testing, we recommend reviewing introduction to statistics textbooks.

[2] This could change in the future if another package is built or {srvyr} is expanded to work with {tidymodels} packages, but no such plans are known at this time.

1. `filter(towny, csd_type == "city")`
2. `towny %>% filter(csd_type == "city")`
3. `towny %>% filter(., csd_type == "city")`
4. `towny %>% filter(.data = ., csd_type == "city")`

Each of these lines of code produces the same output since the argument that takes the dataset is in the first spot in `filter()`. The first two are probably familiar to those who have worked with the tidyverse. The third option functions the same way as the second one but is explicit that `towny` goes into the first argument, and the fourth option indicates that `towny` is going into the named argument of `.data`. Here, we are telling R to take what is on the left side of the pipe (`towny`) and pipe it into the spot with the dot (`.`) — the first argument.

In functions that are not part of the tidyverse, the data argument may not be in the first spot. For example, in `svyttest()`, the data argument is in the second spot, which means we need to place the dot (`.`) in the second spot and not the first. For example:

```
svydata_des %>%
  svyttest(x ~ y, .)
```

By default, the pipe places the left-hand object in the first argument spot. Placing the dot (`.`) in the second argument spot indicates that the survey design object `svydata_des` should be used in the second argument and not the first.

Alternatively, named arguments could be used to place the dot first, as named arguments can appear at any location as in the following:

```
svydata_des %>%
  svyttest(design = ., x ~ y)
```

However, the following code does not work as the `svyttest()` function expects the formula as the first argument when arguments are not named:

```
svydata_des %>%
  svyttest(., x ~ y)
```

6.3 Comparison of proportions and means

We use t-tests to compare two proportions or means. T-tests allow us to determine if one proportion or mean is statistically different from another. They are commonly used to determine if a single estimate differs from a known value (e.g., 0 or 50%) or to compare two group means (e.g., North versus South). Comparing a single estimate to a known value is called a one-sample t-test, and we can set up the hypothesis test as follows:

- $H_0 : \mu = 0$ where μ is the mean outcome and 0 is the value we are comparing it to
- $H_A : \mu \neq 0$

For comparing two estimates, this is called a two-sample t-test. We can set up the hypothesis test as follows:

- $H_0 : \mu_1 = \mu_2$ where μ_i is the mean outcome for group i
- $H_A : \mu_1 \neq \mu_2$

Two-sample t-tests can also be paired or unpaired. If the data come from two different populations (e.g., North versus South), the t-test run is an unpaired or independent samples t-test. Paired t-tests occur when the data come from the same population. This is commonly seen with data from the same population in two different time periods (e.g., before and after an intervention).

The difference between t-tests with non-survey data and survey data is based on the underlying variance estimation difference. Chapter 10 provides a detailed overview of the math behind the mean and sampling error calculations for various sample designs. The functions in the {survey} package account for these nuances, provided the design object is correctly defined.

6.3.1 Syntax

When we do not have survey data, we can use the t.test() function from the {stats} package to run t-tests. This function does not allow for weights or the variance structure that need to be accounted for with survey data. Therefore, we need to use the svyttest() function from {survey} when using survey data. Many of the arguments are the same between the two functions, but there are a few key differences:

- We need to use the survey design object instead of the original data frame
- We can only use a formula and not separate x and y data
- The confidence level cannot be specified and is always set to 95%. However, we show examples of how the confidence level can be changed after running the svyttest() function by using the confint() function.

Here is the syntax for the svyttest() function:

```
svyttest(formula,
         design,
         ...)
```

The arguments are:

- formula: Formula, outcome~group for two-sample, outcome~0 or outcome~1 for one-sample. The group variable must be a factor or character with two levels, or be coded 0/1 or 1/2. We give more details on formula set-up below for different types of tests.
- design: survey design object
- ...: This passes options on for one-sided tests only, and thus, we can specify na.rm=TRUE

Notice that the first argument here is the formula and not the design. This means we must use the dot (.) if we pipe in the survey design object (as described in Section 6.2).

The formula argument can take several different forms depending on what we are measuring. Here are a few common scenarios:

1. One-sample t-test:
   a. Comparison to 0: var ~ 0, where var is the measure of interest, and we compare it to the value 0. For example, we could test if the population mean of household debt is different from 0 given the sample data collected.
   b. Comparison to a different value: var - value ~ 0, where var is the measure of interest and value is what we are comparing to. For example, we could test if the proportion of the population that has blue eyes is different from 25% by using var - 0.25 ~ 0. Note that specifying the formula as var ~ 0.25 is not equivalent and results in a syntax error.
2. Two-sample t-test:
   a. Unpaired:
      - 2 level grouping variable: var ~ groupVar, where var is the measure of interest and groupVar is a variable with two categories. For example, we could test if the average age of the population who voted for president in 2020 differed from the age of people who did not vote. In this case, age

> would be used for var, and a binary variable indicating
> voting activity would be the groupVar.
>
> - 3+ level grouping variable: var ~ groupVar == level,
>   where var is the measure of interest, groupVar is the categor-
>   ical variable, and level is the category level to isolate. For
>   example, we could test if the test scores in one classroom
>   differed from all other classrooms where groupVar would be
>   the variable holding the values for classroom IDs and level
>   is the classroom ID we want to compare to the others.
>
> b. Paired: var_1 - var_2 ~ 0, where var_1 is the first variable of
>   interest and var_2 is the second variable of interest. For example,
>   we could test if test scores on a subject differed between the
>   start and the end of a course, so var_1 would be the test score
>   at the beginning of the course, and var_2 would be the score at
>   the end of the course.

The na.rm argument defaults to FALSE, which means if any data values are
missing, the t-test does not compute. Throughout this chapter, we always set
na.rm = TRUE, but before analyzing the survey data, review the notes provided
in Chapter 11 to better understand how to handle missing data.

Let's walk through a few examples using the RECS data.

6.3.2 Examples

Example 1: One-sample t-test for mean

RECS asks respondents to indicate what temperature they set their house
to during the summer at night[3]. In our data, we have called this variable
SummerTempNight. If we want to see if the average U.S. household sets its
temperature at a value different from $68°F$[4], we could set up the hypothesis
as follows:

- $H_0 : \mu = 68$ where μ is the average temperature U.S. households set their
  thermostat to in the summer at night
- $H_A : \mu \neq 68$

To conduct this in R, we use svyttest() and subtract the temperature on the
left-hand side of the formula:

```
ttest_ex1 <- recs_des %>%
  svyttest(
```

---

[3]Question text: "During the summer, what is your home's typical indoor temperature
inside your home at night?" (U.S. Energy Information Administration, 2020)

[4]This is the temperature that Stephanie prefers at night during the summer, and she
wanted to see if she was different from the population.

```
    formula = SummerTempNight - 68 ~ 0,
    design = .,
    na.rm = TRUE
  )

ttest_ex1
```

```
##
##  Design-based one-sample t-test
##
## data:  SummerTempNight - 68 ~ 0
## t = 85, df = 58, p-value <2e-16
## alternative hypothesis: true mean is not equal to 0
## 95 percent confidence interval:
##   3.288 3.447
## sample estimates:
## mean
## 3.367
```

To pull out specific output, we can use R's built-in \$ operator. For instance, to obtain the estimate $\mu - 68$, we run `ttest_ex1$estimate`.

If we want the average, we take our t-test estimate and add it to 68:

```
ttest_ex1$estimate + 68
```

```
## mean
## 71.37
```

Or, we can use the `survey_mean()` function described in Chapter 5:

```
recs_des %>%
  summarize(mu = survey_mean(SummerTempNight, na.rm = TRUE))
```

```
## # A tibble: 1 x 2
##      mu  mu_se
##   <dbl>  <dbl>
## 1  71.4 0.0397
```

The result is the same in both methods, so we see that the average temperature U.S. households set their thermostat to in the summer at night is 71.4°F. Looking at the output from `svyttest()`, the t-statistic is 84.8, and the p-value is <0.0001, indicating that the average is statistically different from 68°F at an α level of 0.05.

If we want an 80% confidence interval for the test statistic, we can use the function confint() to change the confidence level. Below, we print the default confidence interval (95%), the confidence interval explicitly specifying the level as 95%, and the 80% confidence interval. When the confidence level is 95% either by default or explicitly, R returns a vector with both row and column names. However, when we specify any other confidence level, an unnamed vector is returned, with the first element being the lower bound and the second element being the upper bound of the confidence interval.

```
confint(ttest_ex1)
```

```
##                                       2.5 % 97.5 %
## as.numeric(SummerTempNight - 68) 3.288  3.447
## attr(,"conf.level")
## [1] 0.95
```

```
confint(ttest_ex1, level = 0.95)
```

```
##                                       2.5 % 97.5 %
## as.numeric(SummerTempNight - 68) 3.288  3.447
## attr(,"conf.level")
## [1] 0.95
```

```
confint(ttest_ex1, level = 0.8)
```

```
## [1] 3.316 3.419
## attr(,"conf.level")
## [1] 0.8
```

In this case, neither confidence interval contains 0, and we draw the same conclusion from either that the average temperature households set their thermostat in the summer at night is significantly higher than 68°F.

Example 2: One-sample t-test for proportion

RECS asked respondents if they use air conditioning (A/C) in their home[5]. In our data, we call this variable ACUsed. Let's look at the proportion of U.S. households that use A/C in their homes using the survey_prop() function we learned in Chapter 5.

---

[5]Question text: "Is any air conditioning equipment used in your home?" (U.S. Energy Information Administration, 2020)

```
acprop <- recs_des %>%
  group_by(ACUsed) %>%
  summarize(p = survey_prop())

acprop
```

```
## # A tibble: 2 x 3
##   ACUsed      p   p_se
##   <lgl>   <dbl>  <dbl>
## 1 FALSE   0.113 0.00306
## 2 TRUE    0.887 0.00306
```

Based on this, 88.7% of U.S. households use A/C in their homes. If we wanted to know if this differs from 90%, we could set up our hypothesis as follows:

- $H_0 : p = 0.90$ where p is the proportion of U.S. households that use A/C in their homes
- $H_A : p \neq 0.90$

To conduct this in R, we use the svyttest() function as follows:

```
ttest_ex2 <- recs_des %>%
  svyttest(
    formula = (ACUsed == TRUE) - 0.90 ~ 0,
    design = .,
    na.rm = TRUE
  )

ttest_ex2
```

```
##
##  Design-based one-sample t-test
##
## data:  (ACUsed == TRUE) - 0.9 ~ 0
## t = -4.4, df = 58, p-value = 5e-05
## alternative hypothesis: true mean is not equal to 0
## 95 percent confidence interval:
##  -0.019603 -0.007348
## sample estimates:
##     mean
## -0.01348
```

The output from the svyttest() function can be a bit hard to read. Using the tidy() function from the {broom} package, we can clean up the output into a tibble to more easily understand what the test tells us (Robinson et al., 2023).

```
tidy(ttest_ex2)
```

```
## # A tibble: 1 x 8
##    estimate statistic  p.value parameter conf.low conf.high method
##       <dbl>     <dbl>    <dbl>     <dbl>    <dbl>     <dbl> <chr>
## 1  -0.0135     -4.40 0.0000466        58  -0.0196  -0.00735 Design-ba~
## # i 1 more variable: alternative <chr>
```

The 'tidied' output can also be piped into the {gt} package to create a table ready for publication (see Table 6.1). We go over the {gt} package in Chapter 8. The function `pretty_p_value()` comes from the {prettyunits} package and converts numeric p-values to characters and, by default, prints four decimal places and displays any p-value less than 0.0001 as `"<0.0001"`, though another minimum display p-value can be specified (Csardi, 2023).

```
tidy(ttest_ex2) %>%
  mutate(p.value = pretty_p_value(p.value)) %>%
  gt() %>%
  fmt_number()
```

TABLE 6.1 One-sample t-test output for estimates of U.S. households use A/C in their homes differing from 90%, RECS 2020

estimate	statistic	p.value	parameter	conf.low	conf.high	method	alternative
−0.01	−4.40	<0.0001	58.00	−0.02	−0.01	Design-based one-sample t-test	two.sided

The estimate differs from Example 1 in that it does not display $p - 0.90$ but rather p, or the difference between the U.S. households that use A/C and our comparison proportion. We can see that there is a difference of —1.35 percentage points. Additionally, the t-statistic value in the `statistic` column is —4.4, and the p-value is <0.0001. These results indicate that fewer than 90% of U.S. households use A/C in their homes.

Example 3: Unpaired two-sample t-test

In addition to `ACUsed`, another variable in the RECS data is a household's total electric cost in dollars (`DOLLAREL`).To see if U.S. households with A/C had higher electrical bills than those without, we can set up the hypothesis as follows:

- $H_0 : \mu_{AC} = \mu_{noAC}$ where μ_{AC} is the electrical bill cost for U.S. households that used A/C, and μ_{noAC} is the electrical bill cost for U.S. households that did not use A/C
- $H_A : \mu_{AC} \neq \mu_{noAC}$

Let's take a quick look at the data to see how they are formatted:

```
recs_des %>%
  group_by(ACUsed) %>%
  summarize(mean = survey_mean(DOLLAREL, na.rm = TRUE))
```

```
## # A tibble: 2 x 3
##    ACUsed   mean mean_se
##    <lgl>    <dbl>   <dbl>
## 1 FALSE    1056.    16.0
## 2 TRUE     1422.     5.69
```

To conduct this in R, we use svyttest():

```
ttest_ex3 <- recs_des %>%
  svyttest(
    formula = DOLLAREL ~ ACUsed,
    design = .,
    na.rm = TRUE
  )
```

```
tidy(ttest_ex3) %>%
  mutate(p.value = pretty_p_value(p.value)) %>%
  gt() %>%
  fmt_number()
```

TABLE 6.2 Unpaired two-sample t-test output for estimates of U.S. households electrical bills by A/C use, RECS 2020

estimate	statistic	p.value	parameter	conf.low	conf.high	method	alternative
365.72	21.29	<0.0001	58.00	331.33	400.11	Design-based t-test	two.sided

The results in Table 6.2 indicate that the difference in electrical bills for those who used A/C and those who did not is, on average, \$365.72. The difference appears to be statistically significant as the t-statistic is 21.3 and the p-value is <0.0001. Households that used A/C spent, on average, $365.72 more in 2020 on electricity than households without A/C.

Example 4: Paired two-sample t-test

Let's say we want to test whether the temperature at which U.S. households set their thermostat at night differs depending on the season (comparing summer and winter[6] temperatures). We could set up the hypothesis as follows:

- $H_0 : \mu_{summer} = \mu_{winter}$ where μ_{summer} is the temperature that U.S. households set their thermostat to during summer nights, and μ_{winter} is the temperature that U.S. households set their thermostat to during winter nights
- $H_A : \mu_{summer} \neq \mu_{winter}$

To conduct this in R, we use `svyttest()` by calculating the temperature difference on the left-hand side as follows:

```
ttest_ex4 <- recs_des %>%
  svyttest(
    design = .,
    formula = SummerTempNight - WinterTempNight ~ 0,
    na.rm = TRUE
  )
```

```
tidy(ttest_ex4) %>%
  mutate(p.value = pretty_p_value(p.value)) %>%
  gt() %>%
  fmt_number()
```

The results displayed in Table 6.3 indicate that U.S. households set their thermostat on average 2.9°F warmer in summer nights than winter nights, which is statistically significant (t = 50.8, p-value is <0.0001).

---

[6]Question text: "During the winter, what is your home's typical indoor temperature inside your home at night?" (U.S. Energy Information Administration, 2020)

TABLE 6.3 Paired two-sample t-test output for estimates of U.S. households thermostat temperature by season, RECS 2020

estimate	statistic	p.value	parameter	conf.low	conf.high	method	alternative
2.85	50.83	<0.0001	58.00	2.74	2.96	Design-based one-sample t-test	two.sided

6.4 Chi-squared tests

Chi-squared tests (χ^2) allow us to examine multiple proportions using a goodness-of-fit test, a test of independence, or a test of homogeneity. These three tests have the same χ^2 distributions but with slightly different underlying assumptions.

First, goodness-of-fit tests are used when comparing observed data to expected data. For example, this could be used to determine if respondent demographics (the observed data in the sample) match known population information (the expected data). In this case, we can set up the hypothesis test as follows:

- $H_0 : p_1 = \pi_1,\ p_2 = \pi_2,\ ...,\ p_k = \pi_k$ where p_i is the observed proportion for category i, π_i is the expected proportion for category i, and k is the number of categories
- H_A : at least one level of p_i does not match π_i

Second, tests of independence are used when comparing two types of observed data to see if there is a relationship. For example, this could be used to determine if the proportion of respondents who voted for each political party in the presidential election matches the proportion of respondents who voted for each political party in a local election. In this case, we can set up the hypothesis test as follows:

- H_0 : The two variables/factors are independent
- H_A : The two variables/factors are not independent

Third, tests of homogeneity are used to compare two distributions to see if they match. For example, this could be used to determine if the highest education achieved is the same for both men and women. In this case, we can set up the hypothesis test as follows:

- $H_0 : p_{1a} = p_{1b},\ p_{2a} = p_{2b},\ ...,\ p_{ka} = p_{kb}$ where p_{ia} is the observed proportion of category i for subgroup a, p_{ib} is the observed proportion of category i for subgroup a, and k is the number of categories
- H_A : at least one category of p_{ia} does not match p_{ib}

As with t-tests, the difference between using χ^2 tests with non-survey data and survey data is based on the underlying variance estimation. The functions in the {survey} package account for these nuances, provided the design object is correctly defined. For basic variance estimation formulas for different survey design types, refer to Chapter 10.

6.4.1 Syntax

When we do not have survey data, we may be able to use the `chisq.test()` function from the {stats} package in base R to run chi-squared tests (R Core Team, 2024). However, this function does not allow for weights or the variance structure to be accounted for with survey data. Therefore, when using survey data, we need to use one of two functions:

* `svygofchisq()`: For goodness-of-fit tests
* `svychisq()`: For tests of independence and homogeneity

The non-survey data function of `chisq.test()` requires either a single set of counts and given proportions (for goodness-of-fit tests) or two sets of counts for tests of independence and homogeneity. The functions we use with survey data require respondent-level data and formulas instead of counts. This ensures that the variances are correctly calculated.

First, the function for the goodness-of-fit tests is `svygofchisq()`:

```
svygofchisq(formula,
            p,
            design,
            na.rm = TRUE,
            ...)
```

The arguments are:

* `formula`: Formula specifying a single factor variable
* `p`: Vector of probabilities for the categories of the factor in the correct order. If the probabilities do not sum to 1, they are rescaled to sum to 1.
* `design`: Survey design object
* `...`: Other arguments to pass on, such as `na.rm`

Based on the order of the arguments, we again must use the dot (`.`) notation if we pipe in the survey design object or explicitly name the arguments as described in Section 6.2. For the goodness-of-fit tests, the formula is a single variable `formula = ~var` as we compare the observed data from this variable to the expected data. The expected probabilities are then entered in the `p` argument and need to be a vector of the same length as the number of categories in the variable. For example, if we want to know if the proportion of males and females matches a distribution of 30/70, then the sex variable (with two categories) would be used formula = ~SEX, and the proportions would be included as p = c(.3, .7). It is important to note that the variable entered into the formula should be formatted as either a factor or a character. The examples below provide more detail and tips on how to make sure the levels match up correctly.

For tests of homogeneity and independence, the svychisq() function should be used. The syntax is as follows:

```
svychisq(
  formula,
  design,
  statistic = c("F", "Chisq", "Wald", "adjWald",
                "lincom", "saddlepoint"),
  na.rm = TRUE
)
```

The arguments are:

- formula: Model formula specifying the table (shown in examples)
- design: Survey design object
- statistic: Type of test statistic to use in test (details below)
- na.rm: Remove missing values

There are six statistics that are accepted in this formula. For tests of homogeneity (when comparing cross-tabulations), the F or Chisq statistics should be used[7]. The F statistic is the default and uses the Rao-Scott second-order correction. This correction is designed to assist with complicated sampling designs (i.e., those other than a simple random sample) (Scott, 2007). The Chisq statistic is an adjusted version of the Pearson χ^2 statistic. The version of this statistic in the svychisq() function compares the design effect estimate from the provided survey data to what the χ^2 distribution would have been if the data came from a simple random sampling.

For tests of independence, the Wald and adjWald are recommended as they provide a better adjustment for variable comparisons (Lumley, 2010). If the data have a small number of primary sampling units (PSUs) compared to the degrees of freedom, then the adjWald statistic should be used to account for this. The lincom and saddlepoint statistics are available for more complicated data structures.

The formula argument is always one-sided, unlike the svyttest() function. The two variables of interest should be included with a plus sign: formula = ~ var_1 + var_2. As with the svygofchisq() function, the variables entered into the formula should be formatted as either a factor or a character.

---

[7]These two statistics can also be used for goodness-of-fit tests if the svygofchisq() function is not used.

Additionally, as with the t-test function, both `svygofchisq()` and `svychisq()` have the `na.rm` argument. If any data values are missing, the χ^2 tests assume that `NA` is a category and include it in the calculation. Throughout this chapter, we always set `na.rm = TRUE`, but before analyzing the survey data, review the notes provided in Chapter 11 to better understand how to handle missing data.

6.4.2 Examples

Let's walk through a few examples using the ANES data.

Example 1: Goodness-of-fit test

ANES asked respondents about their highest education level[8]. Based on the data from the 2020 American Community Survey (ACS) 5-year estimates[9], the education distribution of those aged 18+ in the United States (among the 50 states and the District of Columbia) is as follows:

- 11% had less than a high school degree
- 27% had a high school degree
- 29% had some college or an associate's degree
- 33% had a bachelor's degree or higher

If we want to see if the weighted distribution from the ANES 2020 data matches this distribution, we could set up the hypothesis as follows:

- $H_0 : p_1 = 0.11,\ p_2 = 0.27,\ p_3 = 0.29,\ p_4 = 0.33$
- $H_A :$ at least one of the education levels does not match between the ANES and the ACS

To conduct this in R, let's first look at the education variable (`Education`) we have on the ANES data. Using the `survey_mean()` function discussed in Chapter 5, we can see the education levels and estimated proportions.

```
anes_des %>%
  drop_na(Education) %>%
  group_by(Education) %>%
  summarize(p = survey_mean())
```

```
## # A tibble: 5 x 3
##    Education          p    p_se
##    <fct>           <dbl>   <dbl>
## 1 Less than HS   0.0805 0.00568
## 2 High school    0.277  0.0102
```

---

[8]Question text: "What is the highest level of school you have completed or the highest degree you have received?" (American National Election Studies, 2021)

[9]Data was pulled from data.census.gov using the S1501 Education Attainment 2020: ACS 5-Year Estimates Subject Tables.

```
## 3 Post HS        0.290  0.00713
## 4 Bachelor's     0.226  0.00633
## 5 Graduate       0.126  0.00499
```

Based on this output, we can see that we have different levels from the ACS data. Specifically, the education data from ANES include two levels for bachelor's degree or higher (bachelor's and graduate), so these two categories need to be collapsed into a single category to match the ACS data. For this, among other methods, we can use the {forcats} package from the tidyverse (Wickham, 2023). The package's `fct_collapse()` function helps us create a new variable by collapsing categories into a single one. Then, we use the `svygofchisq()` function to compare the ANES data to the ACS data, where we specify the updated design object, the formula using the collapsed education variable, the ACS estimates for education levels as p, and removing `NA` values.

```
anes_des_educ <- anes_des %>%
  mutate(
    Education2 =
      fct_collapse(Education,
        "Bachelor or Higher" = c(
          "Bachelor's",
          "Graduate"
        )
      )
  )

anes_des_educ %>%
  drop_na(Education2) %>%
  group_by(Education2) %>%
  summarize(p = survey_mean())
```

```
## # A tibble: 4 x 3
##   Education2              p    p_se
##   <fct>               <dbl>   <dbl>
## 1 Less than HS       0.0805 0.00568
## 2 High school        0.277  0.0102
## 3 Post HS            0.290  0.00713
## 4 Bachelor or Higher 0.352  0.00732
```

```
chi_ex1 <- anes_des_educ %>%
  svygofchisq(
    formula = ~Education2,
    p = c(0.11, 0.27, 0.29, 0.33),
    design = .,
```

```
    na.rm = TRUE)

chi_ex1
```

```
##
##   Design-based chi-squared test for given probabilities
##
## data:  ~Education2
## X-squared = 2172220, scale = 1.1e+05, df = 2.3e+00, p-value =
## 9e-05
```

The output from the `svygofchisq()` indicates that at least one proportion from ANES does not match the ACS data ($\chi^2 = 2{,}172{,}220$; p-value is <0.0001). To get a better idea of the differences, we can use the `expected` output along with `survey_mean()` to create a comparison table:

```
ex1_table <- anes_des_educ %>%
  drop_na(Education2) %>%
  group_by(Education2) %>%
  summarize(Observed = survey_mean(vartype = "ci")) %>%
  rename(Education = Education2) %>%
  mutate(Expected = c(0.11, 0.27, 0.29, 0.33)) %>%
  select(Education, Expected, everything())

ex1_table
```

```
## # A tibble: 4 x 5
##   Education          Expected Observed Observed_low Observed_upp
##   <fct>                 <dbl>    <dbl>        <dbl>        <dbl>
## 1 Less than HS           0.11   0.0805       0.0691       0.0919
## 2 High school            0.27   0.277        0.257        0.298
## 3 Post HS                0.29   0.290        0.276        0.305
## 4 Bachelor or Higher     0.33   0.352        0.337        0.367
```

This output includes our expected proportions from the ACS that we provided the `svygofchisq()` function along with the output of the observed proportions and their confidence intervals. This table shows that the "high school" and "post HS" categories have nearly identical proportions, but that the other two categories are slightly different. Looking at the confidence intervals, we can see that the ANES data skew to include fewer people in the "less than HS" category and more people in the "bachelor or higher" category. This may be easier to see if we plot this. The code below uses the tabular output to create Figure 6.1.

```
ex1_table %>%
  pivot_longer(
    cols = c("Expected", "Observed"),
    names_to = "Names",
    values_to = "Proportion"
  ) %>%
  mutate(
    Observed_low = if_else(Names == "Observed", Observed_low, NA_real_),
    Observed_upp = if_else(Names == "Observed", Observed_upp, NA_real_),
    Names = if_else(Names == "Observed",
      "ANES (observed)", "ACS (expected)"
    )
  ) %>%
  ggplot(aes(x = Education, y = Proportion, color = Names)) +
  geom_point(alpha = 0.75, size = 2) +
  geom_errorbar(aes(ymin = Observed_low, ymax = Observed_upp),
    width = 0.25
  ) +
  theme_bw() +
  scale_color_manual(name = "Type", values = book_colors[c(4, 1)]) +
  theme(legend.position = "bottom", legend.title = element_blank())
```

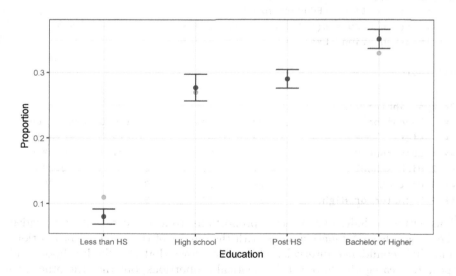

FIGURE 6.1 Expected and observed proportions of education with confidence intervals

Example 2: Test of independence

ANES asked respondents two questions about trust:

- Question text: "How often can you trust the federal government to do what is right?" (American National Election Studies, 2021)
- Question text: "How often can you trust other people?" (American National Election Studies, 2021)

If we want to see if the distributions of these two questions are similar or not, we can conduct a test of independence. Here is how the hypothesis could be set up:

- H_0 : People's trust in the federal government and their trust in other people are independent (i.e., not related)
- H_A : People's trust in the federal government and their trust in other people are not independent (i.e., they are related)

To conduct this in R, we use the svychisq() function to compare the two variables:

```
chi_ex2 <- anes_des %>%
  svychisq(
    formula = ~ TrustGovernment + TrustPeople,
    design = .,
    statistic = "Wald",
    na.rm = TRUE
  )

chi_ex2
```

```
##
##  Design-based Wald test of association
##
## data:  NextMethod()
## F = 21, ndf = 16, ddf = 51, p-value <2e-16
```

The output from svychisq() indicates that the distribution of people's trust in the federal government and their trust in other people are not independent, meaning that they are related. Let's output the distributions in a table to see the relationship. The observed output from the test provides a cross-tabulation of the counts for each category:

```
chi_ex2$observed
```

```
##                           TrustPeople
## TrustGovernment          Always Most of the time About half the time
##    Always                16.470           25.009               31.848
##    Most of the time      11.020          539.377              196.258
##    About half the time   11.772          934.858              861.971
##    Some of the time      17.007         1353.779              839.863
##    Never                  3.174          236.785              174.272
##                           TrustPeople
## TrustGovernment          Some of the time    Never
##    Always                          36.854    5.523
##    Most of the time               206.556   27.184
##    About half the time            428.871   65.024
##    Some of the time               932.628   89.596
##    Never                          217.994  189.307
```

However, we often want to know about the proportions, not just the respondent
counts from the survey. There are a couple of different ways that we can do
this. The first is using the counts from `chi_ex2$observed` to calculate the
proportion. We can then pivot the table to create a cross-tabulation similar to
the counts table above. Adding `group_by()` to the code means that we obtain
the proportions within each variable level. In this case, we are looking at the
distribution of `TrustGovernment` for each level of `TrustPeople`. The resulting
table is shown in Table 6.4.

```
chi_ex2_table <- chi_ex2$observed %>%
  as_tibble() %>%
  group_by(TrustPeople) %>%
  mutate(prop = round(n / sum(n), 3)) %>%
  select(-n) %>%
  pivot_wider(names_from = TrustPeople, values_from = prop) %>%
  gt(rowname_col = "TrustGovernment") %>%
  tab_stubhead(label = "Trust in Government") %>%
  tab_spanner(
    label = "Trust in People",
    columns = everything()
  ) %>%
  cols_label(
    `Most of the time` = md("Most of<br />the time"),
    `About half the time` = md("About half<br />the time"),
    `Some of the time` = md("Some of<br />the time")
  )

chi_ex2_table
```

TABLE 6.4 Proportion of adults in the U.S. by levels of trust in people and government, ANES 2020

Trust in Government	Trust in People				
	Always	Most of the time	About half the time	Some of the time	Never
Always	0.277	0.008	0.015	0.020	0.015
Most of the time	0.185	0.175	0.093	0.113	0.072
About half the time	0.198	0.303	0.410	0.235	0.173
Some of the time	0.286	0.438	0.399	0.512	0.238
Never	0.053	0.077	0.083	0.120	0.503

In Table 6.4, each column sums to 1. For example, we can say that it is estimated that of people who always trust in people, 27.7% also always trust in the government based on the top-left cell, but 5.3% never trust in the government.

The second option is to use the group_by() and survey_mean() functions to calculate the proportions from the ANES design object. Remember that with more than one variable listed in the group_by() statement, the proportions are within the first variable listed. As mentioned above, we are looking at the distribution of TrustGovernment for each level of TrustPeople.

```
chi_ex2_obs <- anes_des %>%
  drop_na(TrustPeople, TrustGovernment) %>%
  group_by(TrustPeople, TrustGovernment) %>%
  summarize(
    Observed = round(survey_mean(vartype = "ci"), 3),
    .groups = "drop"
  )


chi_ex2_obs_table <- chi_ex2_obs %>%
  mutate(prop = paste0(
    Observed, " (", Observed_low, ", ",
    Observed_upp, ")"
  )) %>%
  select(TrustGovernment, TrustPeople, prop) %>%
  pivot_wider(names_from = TrustPeople, values_from = prop) %>%
  gt(rowname_col = "TrustGovernment") %>%
  tab_stubhead(label = "Trust in Government") %>%
  tab_spanner(
    label = "Trust in People",
```

```
    columns = everything()
) %>%
tab_options(page.orientation = "landscape")
```

`chi_ex2_obs_table`

TABLE 6.5 Proportion of adults in the U.S. by levels of trust in people and government with confidence intervals, ANES 2020

Trust in Government	Always	Most of the time	About half the time	Some of the time	Never
Always	0.277 (0.110, 0.444)	0.008 (0.004, 0.012)	0.015 (0.006, 0.024)	0.02 (0.008, 0.033)	0.015 (0.000, 0.029)
Most of the time	0.185 (−0.009, 0.380)	0.175 (0.157, 0.192)	0.093 (0.078, 0.109)	0.113 (0.085, 0.141)	0.072 (0.021, 0.123)
About half the time	0.198 (0.046, 0.350)	0.303 (0.281, 0.324)	0.410 (0.378, 0.441)	0.235 (0.200, 0.271)	0.173 (0.099, 0.246)
Some of the time	0.286 (0.069, 0.503)	0.438 (0.415, 0.462)	0.399 (0.365, 0.433)	0.512 (0.481, 0.543)	0.238 (0.178, 0.298)
Never	0.053 (−0.010, 0.117)	0.077 (0.064, 0.089)	0.083 (0.063, 0.103)	0.120 (0.097, 0.142)	0.503 (0.422, 0.583)

Both methods produce the same output as the svychisq() function. However, calculating the proportions directly from the design object allows us to obtain the variance information. In this case, the output in Table 6.5 displays the survey estimate followed by the confidence intervals. Based on the output, we can see that of those who never trust people, 50.3% also never trust the government, while the proportions of never trusting the government are much lower for each of the other levels of trusting people.

We may find it easier to look at these proportions graphically. We can use ggplot() and facets to provide an overview to create Figure 6.2 below:

```
chi_ex2_obs %>%
  mutate(
    TrustPeople =
      fct_reorder(
        str_c("Trust in People:\n", TrustPeople),
        order(TrustPeople)
      )
  ) %>%
  ggplot(
    aes(x = TrustGovernment, y = Observed, color = TrustGovernment)
  ) +
  facet_wrap(~TrustPeople, ncol = 5) +
  geom_point() +
  geom_errorbar(aes(ymin = Observed_low, ymax = Observed_upp)) +
  ylab("Proportion") +
  xlab("") +
  theme_bw() +
  scale_color_manual(
    name = "Trust in Government",
    values = book_colors
  ) +
  theme(
    axis.text.x = element_blank(),
    axis.ticks.x = element_blank(),
    legend.position = "bottom"
  ) +
  guides(col = guide_legend(nrow = 2))
```

Example 3: Test of homogeneity

Researchers and politicians often look at specific demographics each election cycle to understand how each group is leaning or voting toward candidates. The ANES data are collected post-election, but we can still see if there are differences in how specific demographic groups voted.

If we want to see if there is a difference in how each age group voted for the 2020 candidates, this would be a test of homogeneity, and we can set up the hypothesis as follows:

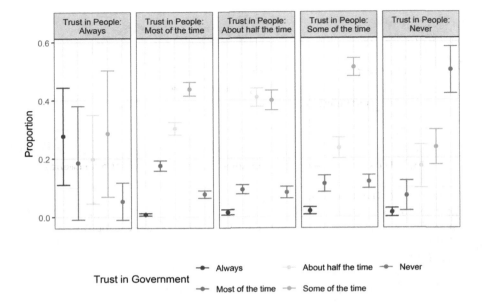

FIGURE 6.2 Proportion of adults in the U.S. by levels of trust in people and government with confidence intervals, ANES 2020

$$H_0 : p_{1_{Biden}} = p_{1_{Trump}} = p_{1_{Other}},$$
$$p_{2_{Biden}} = p_{2_{Trump}} = p_{2_{Other}},$$
$$p_{3_{Biden}} = p_{3_{Trump}} = p_{3_{Other}},$$
$$p_{4_{Biden}} = p_{4_{Trump}} = p_{4_{Other}},$$
$$p_{5_{Biden}} = p_{5_{Trump}} = p_{5_{Other}},$$
$$p_{6_{Biden}} = p_{6_{Trump}} = p_{6_{Other}}$$

where $p_{i_{Biden}}$ is the observed proportion of each age group (i) that voted for Joseph Biden, $p_{i_{Trump}}$ is the observed proportion of each age group (i) that voted for Donald Trump, and $p_{i_{Other}}$ is the observed proportion of each age group (i) that voted for another candidate.

- H_A : at least one category of $p_{i_{Biden}}$ does not match $p_{i_{Trump}}$ or $p_{i_{Other}}$

To conduct this in R, we use the svychisq() function to compare the two variables:

```
chi_ex3 <- anes_des %>%
  drop_na(VotedPres2020_selection, AgeGroup) %>%
  svychisq(
```

```
    formula = ~ AgeGroup + VotedPres2020_selection,
    design = .,
    statistic = "Chisq",
    na.rm = TRUE
  )

chi_ex3
```

```
##
##   Pearson's X^2: Rao & Scott adjustment
##
## data:  NextMethod()
## X-squared = 171, df = 10, p-value <2e-16
```

The output from `svychisq()` indicates a difference in how each age group voted in the 2020 election. To get a better idea of the different distributions, let's output proportions to see the relationship. As we learned in Example 2 above, we can use `chi_ex3$observed`, or if we want to get the variance information (which is crucial with survey data), we can use `survey_mean()`. Remember, when we have two variables in `group_by()`, we obtain the proportions within each level of the variable listed. In this case, we are looking at the distribution of `AgeGroup` for each level of `VotedPres2020_selection`.

```
chi_ex3_obs <- anes_des %>%
  filter(VotedPres2020 == "Yes") %>%
  drop_na(VotedPres2020_selection, AgeGroup) %>%
  group_by(VotedPres2020_selection, AgeGroup) %>%
  summarize(Observed = round(survey_mean(vartype = "ci"), 3))

chi_ex3_obs_table <- chi_ex3_obs %>%
  mutate(prop = paste0(
    Observed, " (", Observed_low, ", ",
    Observed_upp, ")"
  )) %>%
  select(AgeGroup, VotedPres2020_selection, prop) %>%
  pivot_wider(
    names_from = VotedPres2020_selection,
    values_from = prop
  ) %>%
  gt(rowname_col = "AgeGroup") %>%
  tab_stubhead(label = "Age Group")


chi_ex3_obs_table
```

TABLE 6.6 Distribution of age group by presidential candidate selection with confidence intervals

Age Group	Biden	Trump	Other
18–29	0.203 (0.177, 0.229)	0.113 (0.095, 0.132)	0.221 (0.144, 0.298)
30–39	0.168 (0.152, 0.184)	0.146 (0.125, 0.168)	0.302 (0.210, 0.394)
40–49	0.163 (0.146, 0.180)	0.157 (0.137, 0.177)	0.210 (0.130, 0.290)
50–59	0.152 (0.135, 0.170)	0.229 (0.202, 0.256)	0.104 (0.040, 0.168)
60–69	0.177 (0.159, 0.196)	0.193 (0.173, 0.213)	0.103 (0.025, 0.182)
70 or older	0.136 (0.123, 0.149)	0.161 (0.143, 0.179)	0.060 (0.010, 0.109)

In Table 6.6 we can see that the age group distribution that voted for Biden and other candidates was younger than those that voted for Trump. For example, of those who voted for Biden, 20.4% were in the 18–29 age group, compared to only 11.4% of those who voted for Trump were in that age group. Conversely, 23.4% of those who voted for Trump were in the 50–59 age group compared to only 15.4% of those who voted for Biden.

6.5 Exercises

The exercises use the design objects anes_des and recs_des as provided in the Prerequisites box at the beginning of the chapter. Here are some exercises for practicing conducting t-tests using svyttest():

1. Using the RECS data, do more than 50% of U.S. households use A/C (ACUsed)?

2. Using the RECS data, does the average temperature at which U.S. households set their thermostats differ between the day and night in the winter (WinterTempDay and WinterTempNight)?

3. Using the ANES data, does the average age (Age) of those who voted for Joseph Biden in 2020 (VotedPres2020_selection) differ from those who voted for another candidate?

4. If we wanted to determine if the political party affiliation differed for males and females, what test would we use?

   a. Goodness-of-fit test (svygofchisq())
   b. Test of independence (svychisq())
   c. Test of homogeneity (svychisq())

5. In the RECS data, is there a relationship between the type of housing unit (HousingUnitType) and the year the house was built (YearMade)?

6. In the ANES data, is there a difference in the distribution of gender (Gender) across early voting status in 2020 (EarlyVote2020)?

7

Modeling

```
recs_des <- recs_2020 %>%
  as_survey_rep(
    weights = NWEIGHT,
    repweights = NWEIGHT1:NWEIGHT60,
    type = "JK1",
    scale = 59 / 60,
    mse = TRUE
  )
```

7.1   Introduction

Modeling data is a way for researchers to investigate the relationship between a single dependent variable and one or more independent variables. This builds upon the analyses conducted in Chapter 6, which looked at the relationships between just two variables. For example, in Example 3 in Section 6.3.2, we investigated if there is a relationship between the electrical bill cost and whether or not the household used air conditioning (A/C). However, there are potentially other elements that could go into what the cost of electrical bills are in a household (e.g., outside temperature, desired internal temperature, types and number of appliances, etc.).

T-tests only allow us to investigate the relationship of one independent variable at a time, but using models, we can look into multiple variables and even explore interactions between these variables. There are several types of models, but in this chapter, we cover Analysis of Variance (ANOVA) and linear regression models following common normal (Gaussian) and logit models. Jonas Kristoffer Lindeløv has an interesting discussion[1] of many statistical tests and models being equivalent to a linear model. For example, a one-way ANOVA is a linear model with one categorical independent variable, and a two-sample t-test is an ANOVA where the independent variable has exactly two levels.

When modeling data, it is helpful to first create an equation that provides an overview of what we are modeling. The main structure of these models is as follows:

$$y_i = \beta_0 + \sum_{i=1}^{p} \beta_i x_i + \epsilon_i$$

---

[1]https://lindeloev.github.io/tests-as-linear/

where y_i is the outcome, β_0 is an intercept, $x_1, \cdots, x_p$ are the predictors with $\beta_1, \cdots, \beta_p$ as the associated coefficients, and ϵ_i is the error. Not all models have all components. For example, some models may not include an intercept (β_0), may have interactions between different independent variables (x_i), or may have different underlying structures for the dependent variable (y_i). However, all linear models have the independent variables related to the dependent variable in a linear form.

To specify these models in R, the formulas are the same with both survey data and other data. The left side of the formula is the response/dependent variable, and the right side has the predictor/independent variable(s). There are many symbols used in R to specify the formula.

For example, a linear formula mathematically notated as

$$y_i = \beta_0 + \beta_1 x_i + \epsilon_i$$

would be specified in R as y~x where the intercept is not explicitly included. To fit a model with no intercept, that is,

$$y_i = \beta_1 x_i + \epsilon_i$$

it can be specified in R as y~x-1. Formula notation details in R can be found in the help file for formula[2]. A quick overview of the common formula notation is in Table 7.1:

TABLE 7.1 Common symbols in formula notation

Symbol	Example	Meaning
+	+x	include this variable
-	-x	delete this variable
:	x:z	include the interaction between these variables
*	x*z	include these variables and the interactions between them
^n	(x+y+z)^3	include these variables and all interactions up to n-way
I	I(x-z)	as-is: include a new variable that is calculated inside the parentheses (e.g., x-z, x*z, x/z are possible calculations that could be done)

There are often multiple ways to specify the same formula. For example, consider the following equation using the mtcars dataset that is built into R:

$$mpg_i = \beta_0 + \beta_1 cyl_i + \beta_2 disp_i + \beta_3 hp_i + \beta_4 cyl_i disp_i + \beta_5 cyl_i hp_i + \beta_6 disp_i hp_i + \epsilon_i$$

---

[2]Use help(formula) or ?formula in R

This could be specified in R code as any of the following:

- `mpg ~ (cyl + disp + hp)^2`
- `mpg ~ cyl + disp + hp + cyl:disp + cyl:hp + disp:hp`
- `mpg ~ cyl*disp + cyl*hp + disp*hp`

In the above options, the ways the : and * notations are implemented are different. Using : only includes the interactions and not the main effects, while using * includes the main effects and all possible interactions. Table 7.2 provides an overview of the syntax and differences between the two notations.

TABLE 7.2 Differences in formulas for : and * code syntax

Symbol	Syntax	Formula
:	`mpg ~ cyl:disp:hp`	$mpg_i = \beta_0 + \beta_4 cyl_i disp_i + \beta_5 cyl_i hp_i + \beta_6 disp_i hp_i + \epsilon_i$
*	`mpg ~ cyl*disp*hp`	$mpg_i = \beta_0 + \beta_1 cyl_i + \beta_2 disp_i + \beta_3 hp_i + \beta_4 cyl_i disp_i + \beta_5 cyl_i hp_i + \beta_6 disp_i hp_i + \beta_7 cyl_i disp_i hp_i + \epsilon_i$

When using non-survey data, such as experimental or observational data, researchers use the `glm()` function for linear models. With survey data, however, we use `svyglm()` from the {survey} package to ensure that we account for the survey design and weights in modeling[3]. This allows us to generalize a model to the population of interest and accounts for the fact that the observations in the survey data may not be independent. As discussed in Chapter 6, modeling survey data cannot be directly done in {srvyr}, but can be done in the {survey} package (Lumley, 2010). In this chapter, we provide syntax and examples for linear models, including ANOVA, normal linear regression, and logistic regression. For details on other types of regression, including ordinal regression, log-linear models, and survival analysis, refer to Lumley (2010). Lumley (2010) also discusses custom models such as a negative binomial or Poisson model in appendix E of his book.

7.2   Analysis of variance

In ANOVA, we are testing whether the mean of an outcome is the same across two or more groups. Statistically, we set up this as follows:

---

[3]There is some debate about whether weights should be used in regression (Bollen et al., 2016; Gelman, 2007). However, for the purposes of providing complete information on how to analyze complex survey data, this chapter includes weights.

- $H_0 : \mu_1 = \mu_2 = \cdots = \mu_k$ where μ_i is the mean outcome for group i
- H_A : At least one mean is different

An ANOVA test is also a linear model, we can re-frame the problem using the framework as:

$$y_i = \sum_{i=1}^{k} \mu_i x_i + \epsilon_i$$

where x_i is a group indicator for groups $1, \cdots, k$.

Some assumptions when using ANOVA on survey data include:

- The outcome variable is normally distributed within each group.
- The variances of the outcome variable between each group are approximately equal.
- We do NOT assume independence between the groups as with ANOVA on non-survey data. The covariance is accounted for in the survey design.

7.2.1 Syntax

To perform this type of analysis in R, the general syntax is as follows:

```
des_obj %>%
  svyglm(
    formula = outcome ~ group,
    design = .,
    na.action = na.omit,
    df.resid = NULL
  )
```

The arguments are:

- `formula`: formula in the form of `outcome~group`. The group variable must be a factor or character.
- `design`: a `tbl_svy` object created by `as_survey`
- `na.action`: handling of missing data
- `df.resid`: degrees of freedom for Wald tests (optional); defaults to using `degf(design)-(g-1)` where g is the number of groups

The function `svyglm()` does not have the design as the first argument so the dot (.) notation is used to pass it with a pipe (see Chapter 6 for more details). The default for missing data is `na.omit`. This means that we are removing all records with any missing data in either predictors or outcomes from analyses. There are other options for handling missing data, and we recommend looking at the help documentation for `na.omit` (run `help(na.omit)` or `?na.omit`) for more information on options to use for `na.action`. For a discussion on how to handle missing data, see Chapter 11.

7.2.2 Example

Looking at an example helps us discuss the output and how to interpret the results. In RECS, respondents are asked what temperature they set their thermostat to during the evening when using A/C during the summer[4]. To analyze these data, we filter the respondents to only those using A/C (`ACUsed`)[5]. Then, if we want to see if there are regional differences, we can use `group_by()`. A descriptive analysis of the temperature at night (`SummerTempNight`) set by region and the sample sizes is displayed below.

```
recs_des %>%
  filter(ACUsed) %>%
  group_by(Region) %>%
  summarize(
    SMN = survey_mean(SummerTempNight, na.rm = TRUE),
    n = unweighted(n()),
    n_na = unweighted(sum(is.na(SummerTempNight)))
  )
```

```
## # A tibble: 4 x 5
##   Region        SMN SMN_se     n  n_na
##   <fct>       <dbl>  <dbl> <int> <int>
## 1 Northeast    69.7  0.103  3204     0
## 2 Midwest      71.0  0.0897 3619     0
## 3 South        71.8  0.0536 6065     0
## 4 West         72.5  0.129  3283     0
```

In the following code, we test whether this temperature varies by region by first using `svyglm()` to run the test and then using `broom::tidy()` to display the output. Note that the temperature setting is set to NA when the household does not use A/C, and since the default handling of NAs is `na.action=na.omit`, records that do not use A/C are not included in this regression.

```
anova_out <- recs_des %>%
  svyglm(
    design = .,
```

---

[4]Question text: "During the summer, what is your home's typical indoor temperature inside your home at night?" (U.S. Energy Information Administration, 2020)

[5]Question text: "Is any air conditioning equipment used in your home?" (U.S. Energy Information Administration, 2020)

```
  formula = SummerTempNight ~ Region)

tidy(anova_out)

## # A tibble: 4 x 5
##   term          estimate std.error statistic    p.value
##   <chr>            <dbl>     <dbl>     <dbl>      <dbl>
## 1 (Intercept)       69.7     0.103     674.   3.69e-111
## 2 RegionMidwest      1.34    0.138       9.68 1.46e- 13
## 3 RegionSouth        2.05    0.128      16.0  1.36e- 22
## 4 RegionWest         2.80    0.177      15.9  2.27e- 22
```

In the output above, we can see the estimated coefficients (`estimate`), estimated standard errors of the coefficients (`std.error`), the t-statistic (`statistic`), and the p-value for each coefficient. In this output, the intercept represents the reference value of the Northeast region. The other coefficients indicate the difference in temperature relative to the Northeast region. For example, in the Midwest, temperatures are set, on average, 1.34 (p-value is <0.0001) degrees higher than in the Northeast during summer nights, and each region sets its thermostats at significantly higher temperatures than the Northeast.

If we wanted to change the reference value, we would reorder the factor before modeling using the `relevel()` function from {stats} or using one of many factor ordering functions in {forcats} such as `fct_relevel()` or `fct_infreq()`. For example, if we wanted the reference level to be the Midwest region, we could use the following code with the results in Table 7.3. Note the usage of the `gt()` function on top of `tidy()` to print a nice-looking output table (Iannone et al., 2024; Robinson et al., 2023) (see Chapter 8 for more information on the {gt} package).

```
anova_out_relevel <- recs_des %>%
  mutate(Region = fct_relevel(Region, "Midwest", after = 0)) %>%
  svyglm(
    design = .,
    formula = SummerTempNight ~ Region
  )

tidy(anova_out_relevel) %>%
  mutate(p.value = pretty_p_value(p.value)) %>%
  gt() %>%
  fmt_number()
```

TABLE 7.3 ANOVA output for estimates of thermostat temperature setting at night by region with Midwest as the reference region, RECS 2020

term	estimate	std.error	statistic	p.value
(Intercept)	71.04	0.09	791.83	<0.0001
RegionNortheast	−1.34	0.14	−9.68	<0.0001
RegionSouth	0.71	0.10	6.91	<0.0001
RegionWest	1.47	0.16	9.17	<0.0001

This output now has the coefficients indicating the difference in temperature relative to the Midwest region. For example, in the Northeast, temperatures are set, on average, 1.34 (p-value is <0.0001) degrees lower than in the Midwest during summer nights, and each region sets its thermostats at significantly lower temperatures than the Midwest. This is the reverse of what we saw in the prior model, as we are still comparing the same two regions, just from different reference points.

7.3 Normal linear regression

Normal linear regression is a more generalized method than ANOVA, where we fit a model of a continuous outcome with any number of categorical or continuous predictors (whereas ANOVA only has categorical predictors) and is similarly specified as:

$$y_i = \beta_0 + \sum_{i=1}^{p} \beta_i x_i + \epsilon_i \tag{7.1}$$

where y_i is the outcome, β_0 is an intercept, $x_1, \cdots, x_p$ are the predictors with $\beta_1, \cdots, \beta_p$ as the associated coefficients, and ϵ_i is the error.

Assumptions in normal linear regression using survey data include:

- The residuals (ϵ_i) are normally distributed, but there is not an assumption of independence, and the correlation structure is captured in the survey design object
- There is a linear relationship between the outcome variable and the independent variables
- The residuals are homoscedastic; that is, the error term is the same across all values of independent variables

7.3.1 Syntax

The syntax for this regression uses the same function as ANOVA but can have more than one variable listed on the right-hand side of the formula:

```
des_obj %>%
  svyglm(
    formula = outcomevar ~ x1 + x2 + x3,
    design = .,
    na.action = na.omit,
    df.resid = NULL
  )
```

The arguments are:

- `formula`: formula in the form of `y~x`
- `design`: a `tbl_svy` object created by `as_survey`
- `na.action`: handling of missing data
- `df.resid`: degrees of freedom for Wald tests (optional); defaults to using `degf(design)-p` where p is the rank of the design matrix

As discussed in Section 7.1, the formula on the right-hand side can be specified in many ways, for example, denoting whether or not interactions are desired.

7.3.2 Examples

Example 1: Linear regression with a single variable

On RECS, we can obtain information on the square footage of homes[6] and the electric bills. We assume that square footage is related to the amount of money spent on electricity and examine a model for this. Before any modeling, we first plot the data to determine whether it is reasonable to assume a linear relationship. In Figure 7.1, each hexagon represents the weighted count of households in the bin, and we can see a general positive linear trend (as the square footage increases, so does the amount of money spent on electricity).

```
recs_2020 %>%
  ggplot(aes(
    x = TOTSQFT_EN,
    y = DOLLAREL,
    weight = NWEIGHT / 1000000
  )) +
  geom_hex() +
```

---

[6]Question text: "What is the square footage of your home?" (U.S. Energy Information Administration, 2020)

```
scale_fill_gradientn(
  guide = "colorbar",
  name = "Housing Units\n(Millions)",
  labels = scales::comma,
  colors = book_colors[c(3, 2, 1)]
) +
xlab("Total square footage") +
ylab("Amount spent on electricity") +
scale_y_continuous(labels = scales::dollar_format()) +
scale_x_continuous(labels = scales::comma_format()) +
theme_minimal()
```

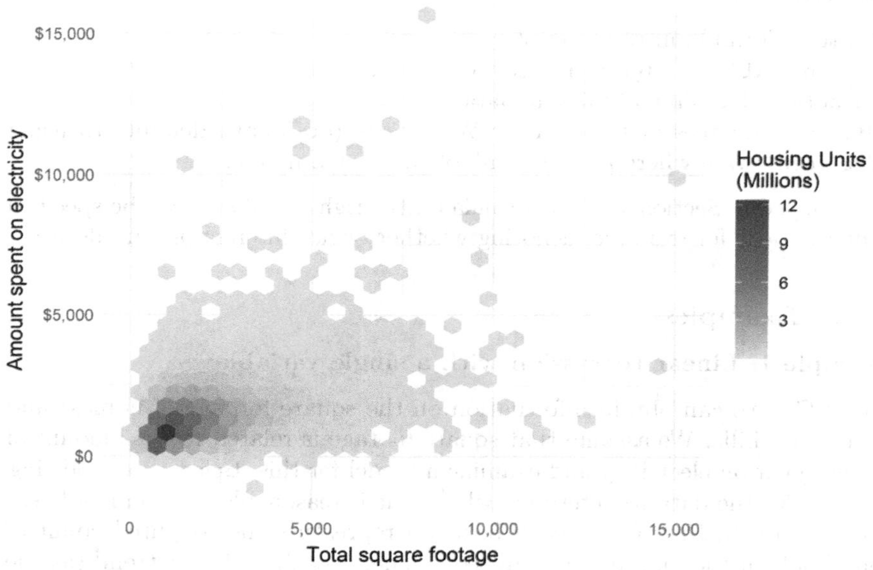

FIGURE 7.1 Relationship between square footage and dollars spent on electricity, RECS 2020

Given that the plot shows a potentially increasing relationship between square footage and electricity expenditure, fitting a model allows us to determine if the relationship is statistically significant. The model is fit below with electricity expenditure as the outcome.

```
m_electric_sqft <- recs_des %>%
  svyglm(
    design = .,
```

```
    formula = DOLLAREL ~ TOTSQFT_EN,
    na.action = na.omit
)
```

```
tidy(m_electric_sqft) %>%
  mutate(p.value = pretty_p_value(p.value)) %>%
  gt() %>%
  fmt_number()
```

TABLE 7.4 Linear regression output predicting electricity expenditure given square footage, RECS 2020

term	estimate	std.error	statistic	p.value
(Intercept)	836.72	12.77	65.51	<0.0001
TOTSQFT_EN	0.30	0.01	41.67	<0.0001

In Table 7.4, we can see the estimated coefficients (estimate), estimated standard errors of the coefficients (std.error), the t-statistic (statistic), and the p-value for each coefficient. In these results, we can say that, on average, for every additional square foot of house size, the electricity bill increases by 30 cents, and that square footage is significantly associated with electricity expenditure (p-value is <0.0001).

This is a straightforward model, and there are likely many more factors related to electricity expenditure, including the type of cooling, number of appliances, location, and more. However, starting with one-variable models can help analysts understand what potential relationships there are between variables before fitting more complex models. Often, we start with known relationships before building models to determine what impact additional variables have on the model.

Example 2: Linear regression with multiple variables and interactions

In the following example, a model is fit to predict electricity expenditure, including census region (factor/categorical), urbanicity (factor/categorical), square footage (double/numeric), and whether A/C is used (logical/categorical) with all two-way interactions also included. In this example, we are choosing to fit this model without an intercept (using -1 in the formula). This results in an intercept estimate for each region instead of a single intercept for all data.

```
m_electric_multi <- recs_des %>%
  svyglm(
```

```
    design = .,
    formula =
      DOLLAREL ~ (Region + Urbanicity + TOTSQFT_EN + ACUsed)^2 - 1,
    na.action = na.omit
  )
```

```
tidy(m_electric_multi) %>%
  mutate(p.value = pretty_p_value(p.value)) %>%
  gt() %>%
  fmt_number()
```

As shown in Table 7.5, there are many terms in this model. To test whether coefficients for a term are different from zero, the `regTermTest()` function can be used. For example, in the above regression, we can test whether the interaction of region and urbanicity is significant as follows:

```
urb_reg_test <- regTermTest(m_electric_multi, ~ Urbanicity:Region)
urb_reg_test
```

```
## Wald test for Urbanicity:Region
##  in svyglm(design = ., formula = DOLLAREL ~ (Region + Urbanicity +
##      TOTSQFT_EN + ACUsed)^2 - 1, na.action = na.omit)
## F =  6.851  on  6  and  35  df: p= 7.2e-05
```

This output indicates there is a significant interaction between urbanicity and region (p-value is <0.0001).

To examine the predictions, residuals, and more from the model, the `augment()` function from {broom} can be used. The `augment()` function returns a tibble with the independent and dependent variables and other fit statistics. The `augment()` function has not been specifically written for objects of class `svyglm`, and as such, a warning is displayed indicating this at this time. As it was not written exactly for this class of objects, a little tweaking needs to be done after using `augment()`. To obtain the standard error of the predicted values (`.se.fit`), we need to use the `attr()` function on the predicted values (`.fitted`) created by `augment()`. Additionally, the predicted values created are outputted with a type of `svrep`. If we want to plot the predicted values, we need to use `as.numeric()` to get the predicted values into a numeric format to work with. However, it is important to note that this adjustment must be completed after the standard error adjustment.

TABLE 7.5 Linear regression output predicting electricity expenditure given region, urbanicity, square footage, A/C usage, and one-way interactions, RECS 2020

term	estimate	std.error	statistic	p.value
RegionNortheast	543.73	56.57	9.61	<0.0001
RegionMidwest	702.16	78.12	8.99	<0.0001
RegionSouth	938.74	46.99	19.98	<0.0001
RegionWest	603.27	36.31	16.61	<0.0001
UrbanicityUrban Cluster	73.03	81.50	0.90	0.3764
UrbanicityRural	204.13	80.69	2.53	0.0161
TOTSQFT_EN	0.24	0.03	8.65	<0.0001
ACUsedTRUE	252.06	54.05	4.66	<0.0001
RegionMidwest: UrbanicityUrban Cluster	183.06	82.38	2.22	0.0328
RegionSouth: UrbanicityUrban Cluster	152.56	76.03	2.01	0.0526
RegionWest: UrbanicityUrban Cluster	98.02	75.16	1.30	0.2007
RegionMidwest: UrbanicityRural	312.83	50.88	6.15	<0.0001
RegionSouth: UrbanicityRural	220.00	55.00	4.00	0.0003
RegionWest: UrbanicityRural	180.97	58.70	3.08	0.0040
RegionMidwest: TOTSQFT_EN	−0.05	0.02	−2.09	0.0441
RegionSouth: TOTSQFT_EN	0.00	0.03	0.11	0.9109
RegionWest: TOTSQFT_EN	−0.03	0.03	−1.00	0.3254
RegionMidwest: ACUsedTRUE	−292.97	60.24	−4.86	<0.0001
RegionSouth: ACUsedTRUE	−294.07	57.44	−5.12	<0.0001
RegionWest: ACUsedTRUE	−77.68	47.05	−1.65	0.1076
UrbanicityUrban Cluster: TOTSQFT_EN	−0.04	0.02	−1.63	0.1112
UrbanicityRural: TOTSQFT_EN	−0.06	0.02	−2.60	0.0137
UrbanicityUrban Cluster: ACUsedTRUE	−130.23	60.30	−2.16	0.0377
UrbanicityRural: ACUsedTRUE	−33.80	59.30	−0.57	0.5724
TOTSQFT_EN: ACUsedTRUE	0.08	0.02	3.48	0.0014

```
fitstats <-
  augment(m_electric_multi) %>%
  mutate(
```

```
    .se.fit = sqrt(attr(.fitted, "var")),
    .fitted = as.numeric(.fitted)
)
```

```
fitstats
```

```
## # A tibble: 18,496 x 13
##    DOLLAREL Region    Urbanicity TOTSQFT_EN ACUsed `(weights)` .fitted
##       <dbl> <fct>     <fct>           <dbl> <lgl>        <dbl>   <dbl>
##  1    1955. West      Urban Area       2100 TRUE         0.492   1397.
##  2     713. South     Urban Area        590 TRUE         1.35    1090.
##  3     335. West      Urban Area        900 TRUE         0.849   1043.
##  4    1425. South     Urban Area       2100 TRUE         0.793   1584.
##  5    1087  Northeast Urban Area        800 TRUE         1.49    1055.
##  6    1896. South     Urban Area       4520 TRUE         1.09    2375.
##  7    1418. South     Urban Area       2100 TRUE         0.851   1584.
##  8    1237. South     Urban Clu~        900 FALSE        1.45    1349.
##  9     538. South     Urban Area        750 TRUE         0.185   1142.
## 10     625. West      Urban Area        760 TRUE         1.06    1002.
## # i 18,486 more rows
## # i 6 more variables: .resid <dbl>, .hat <dbl>, .sigma <dbl>,
## #   .cooksd <dbl>, .std.resid <dbl>, .se.fit <dbl>
```

These results can then be used in a variety of ways, including examining residual plots as illustrated in the code below and Figure 7.2. In the residual plot, we look for any patterns in the data. If we do see patterns, this may indicate a violation of the heteroscedasticity assumption and the standard errors of the coefficients may be incorrect. In Figure 7.2, we do not see a strong pattern indicating that our assumption of heteroscedasticity may hold.

```
fitstats %>%
  ggplot(aes(x = .fitted, .resid)) +
  geom_point(alpha = .1) +
  geom_hline(yintercept = 0, color = "red") +
  theme_minimal() +
  xlab("Fitted value of electricity cost") +
  ylab("Residual of model") +
  scale_y_continuous(labels = scales::dollar_format()) +
  scale_x_continuous(labels = scales::dollar_format())
```

Additionally, augment() can be used to predict outcomes for data not used in modeling. Perhaps we would like to predict the energy expenditure for a home in an urban area in the south that uses A/C and is 2,500 square feet. To do this, we first make a tibble including that additional data and then use the newdata argument in the augment() function. As before, to obtain the standard error of the predicted values, we need to use the attr() function.

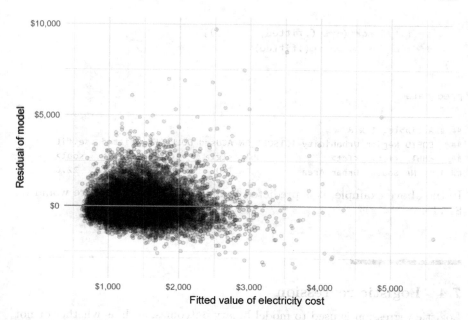

FIGURE 7.2 Residual plot of electric cost model with the following covariates: Region, Urbanicity, TOTSQFT_EN, and ACUsed

```
add_data <- recs_2020 %>%
  select(
    DOEID, Region, Urbanicity,
    TOTSQFT_EN, ACUsed,
    DOLLAREL
  ) %>%
  rbind(
    tibble(
      DOEID = NA,
      Region = "South",
      Urbanicity = "Urban Area",
      TOTSQFT_EN = 2500,
      ACUsed = TRUE,
      DOLLAREL = NA
    )
  ) %>%
  tail(1)

pred_data <- augment(m_electric_multi, newdata = add_data) %>%
  mutate(
```

```
    .se.fit = sqrt(attr(.fitted, "var")),
    .fitted = as.numeric(.fitted)
  )
```

```
pred_data
```

```
## # A tibble: 1 x 8
##   DOEID Region Urbanicity TOTSQFT_EN ACUsed DOLLAREL .fitted .se.fit
##   <dbl> <fct>  <fct>           <dbl> <lgl>     <dbl>   <dbl>   <dbl>
## 1    NA South  Urban Area       2500 TRUE         NA   1715.    22.6
```

In the above example, it is predicted that the energy expenditure would be $1,715.

7.4 Logistic regression

Logistic regression is used to model binary outcomes, such as whether or not someone voted. There are several instances where an outcome may not be originally binary but is collapsed into being binary. For example, given that gender is often asked in surveys with multiple response options and not a binary scale, many researchers now code gender in logistic modeling as "cis-male" compared to not "cis-male." We could also convert a 4-point Likert scale that has levels of "Strongly Agree," "Agree," "Disagree," and "Strongly Disagree" to group the agreement levels into one group and disagreement levels into a second group.

Logistic regression is a specific case of the generalized linear model (GLM). A GLM uses a link function to link the response variable to the linear model. If we tried to use a normal linear regression with a binary outcome, many assumptions would not hold, namely, the response would not be continuous. Logistic regression allows us to link a linear model between the covariates and the propensity of an outcome. In logistic regression, the link model is the logit function. Specifically, the model is specified as follows:

$$y_i \sim \text{Bernoulli}(\pi_i)$$

$$\log\left(\frac{\pi_i}{1 - \pi_i}\right) = \beta_0 + \sum_{i=1}^{n} \beta_i x_i \qquad (7.2)$$

which can be re-expressed as

$$\pi_i = \frac{\exp\left(\beta_0 + \sum_{i=1}^{n} \beta_i x_i\right)}{1 + \exp\left(\beta_0 + \sum_{i=1}^{n} \beta_i x_i\right)}$$

where y_i is the outcome, β_0 is an intercept, and $x_1, \cdots, x_n$ are the predictors with $\beta_1, \cdots, \beta_n$ as the associated coefficients.

The Bernoulli distribution is a distribution which has an outcome of 0 or 1 given some probability (π_i) in this case, and we model π_i as a function of the covariates x_i using this logit link.

Assumptions in logistic regression using survey data include:

- The outcome variable has two levels
- There is a linear relationship between the independent variables and the log odds (the equation for the logit function)
- The residuals are homoscedastic; that is, the error term is the same across all values of independent variables

7.4.1 Syntax

The syntax for logistic regression is as follows:

```
des_obj %>%
  svyglm(
    formula = outcomevar ~ x1 + x2 + x3,
    design = .,
    na.action = na.omit,
    df.resid = NULL,
    family = quasibinomial
  )
```

The arguments are:

- `formula`: Formula in the form of `y~x`
- `design`: a tbl_svy object created by `as_survey`
- `na.action`: handling of missing data
- `df.resid`: degrees of freedom for Wald tests (optional); defaults to using `degf(design)-p` where p is the rank of the design matrix
- `family`: the error distribution/link function to be used in the model

Note `svyglm()` is the same function used in both ANOVA and normal linear regression. However, we've added the link function quasibinomial. While we can use the binomial link function, it is recommended to use the quasibinomial as our weights may not be integers, and the quasibinomial also allows for overdispersion (Lumley, 2010; McCullagh and Nelder, 1989; R Core Team, 2024). The quasibinomial family has a default logit link, which is specified in the equations above. When specifying the outcome variable, it is likely specified in one of three ways with survey data:

- A two-level factor variable where the first level of the factor indicates a "failure," and the second level indicates a "success"
- A numeric variable which is 1 or 0 where 1 indicates a success
- A logical variable where TRUE indicates a success

7.4.2 Examples

Example 1: Logistic regression with single variable

In the following example, we use the ANES data to model whether someone usually has trust in the government[7] by whom someone voted for president in 2020. As a reminder, the leading candidates were Biden and Trump, though people could vote for someone else not in the Democratic or Republican parties. Those votes are all grouped into an "Other" category. We first create a binary outcome for trusting in the government by collapsing "Always" and "Most of the time" into a single-factor level, and the other response options ("About half the time," "Some of the time," and "Never") into a second factor level. Next, a scatter plot of the raw data is not useful, as it is all 0 and 1 outcomes; so instead, we plot a summary of the data.

```
anes_des_der <- anes_des %>%
  mutate(TrustGovernmentUsually = case_when(
    is.na(TrustGovernment) ~ NA,
    TRUE ~ TrustGovernment %in% c("Always", "Most of the time")
  ))

anes_des_der %>%
  group_by(VotedPres2020_selection) %>%
  summarize(
    pct_trust = survey_mean(TrustGovernmentUsually,
      na.rm = TRUE,
      proportion = TRUE,
      vartype = "ci"
    ),
    .groups = "drop"
  ) %>%
  filter(complete.cases(.)) %>%
  ggplot(aes(
    x = VotedPres2020_selection, y = pct_trust,
    fill = VotedPres2020_selection
```

---

[7]Question text: "How often can you trust the federal government in Washington to do what is right?" (American National Election Studies, 2021)

```
)) +
geom_bar(stat = "identity") +
geom_errorbar(aes(ymin = pct_trust_low, ymax = pct_trust_upp),
  width = .2
) +
scale_fill_manual(values = c("#0b3954", "#bfd7ea", "#8d6b94")) +
xlab("Election choice (2020)") +
ylab("Usually trust the government") +
scale_y_continuous(labels = scales::percent) +
guides(fill = "none") +
theme_minimal()
```

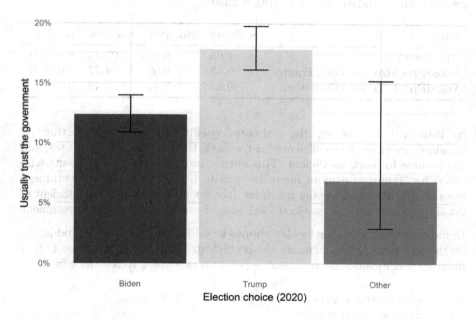

FIGURE 7.3 Relationship between candidate selection and trust in government, ANES 2020

Looking at Figure 7.3, it appears that people who voted for Trump are more likely to say that they usually have trust in the government compared to those who voted for Biden and other candidates. To determine if this insight is accurate, we next fit the model.

```
logistic_trust_vote <- anes_des_der %>%
  svyglm(
    design = .,
```

```
    formula = TrustGovernmentUsually ~ VotedPres2020_selection,
    family = quasibinomial
)
```

```
tidy(logistic_trust_vote) %>%
  mutate(p.value = pretty_p_value(p.value)) %>%
  gt() %>%
  fmt_number()
```

TABLE 7.6 Logistic regression output predicting trust in government by presidential candidate selection, RECS 2020

term	estimate	std.error	statistic	p.value
(Intercept)	−1.96	0.07	−27.45	<0.0001
VotedPres2020_selectionTrump	0.43	0.09	4.72	<0.0001
VotedPres2020_selectionOther	−0.65	0.44	−1.49	0.1429

In Table 7.6, we can see the estimated coefficients (`estimate`), estimated standard errors of the coefficients (`std.error`), the t-statistic (`statistic`), and the p-value for each coefficient. This output indicates that respondents who voted for Trump are more likely to usually have trust in the government compared to those who voted for Biden (the reference level). The coefficient of 0.435 represents the increase in the log odds of usually trusting the government.

In most cases, it is easier to talk about the odds instead of the log odds. To do this, we need to exponentiate the coefficients. We can use the same `tidy()` function but include the argument `exponentiate = TRUE` to see the odds.

```
tidy(logistic_trust_vote, exponentiate = TRUE) %>%
  select(term, estimate) %>%
  gt() %>%
  fmt_number()
```

Using the output in Table 7.7, we can interpret this as saying that the odds of usually trusting the government for someone who voted for Trump is 154% as likely to trust the government compared to a person who voted for Biden (the reference level). In comparison, a person who voted for neither Biden nor Trump is 52% as likely to trust the government as someone who voted for Biden.

As with linear regression, the `augment()` can be used to predict values. By default, the prediction is the link function, not the probability model. To predict

TABLE 7.7 Logistic regression predicting trust in government by presidential candidate selection with exponentiated coefficients (odds), RECS 2020

term	estimate
(Intercept)	0.14
VotedPres2020_selectionTrump	1.54
VotedPres2020_selectionOther	0.52

the probability, add an argument of `type.predict="response"` as demonstrated below:

```
logistic_trust_vote %>%
  augment(type.predict = "response") %>%
  mutate(
    .se.fit = sqrt(attr(.fitted, "var")),
    .fitted = as.numeric(.fitted)
  ) %>%
  select(
    TrustGovernmentUsually,
    VotedPres2020_selection,
    .fitted,
    .se.fit
  )
```

```
## # A tibble: 6,212 x 4
##    TrustGovernmentUsually VotedPres2020_selection .fitted .se.fit
##    <lgl>                  <fct>                      <dbl>   <dbl>
##  1 FALSE                  Other                     0.0681  0.0279
##  2 FALSE                  Biden                     0.123   0.00772
##  3 FALSE                  Biden                     0.123   0.00772
##  4 FALSE                  Trump                     0.178   0.00919
##  5 FALSE                  Biden                     0.123   0.00772
##  6 FALSE                  Trump                     0.178   0.00919
##  7 FALSE                  Biden                     0.123   0.00772
##  8 FALSE                  Biden                     0.123   0.00772
##  9 TRUE                   Biden                     0.123   0.00772
## 10 FALSE                  Biden                     0.123   0.00772
## # i 6,202 more rows
```

Example 2: Interaction effects

Let's look at another example with interaction effects. If we're interested in understanding the demographics of people who voted for Biden among all voters in 2020, we could include the indicator of whether respondents voted early (`EarlyVote2020`) and their income group (`Income7`) in our model.

First, we need to subset the data to 2020 voters and then create an indicator for who voted for Biden.

```
anes_des_ind <- anes_des %>%
  filter(!is.na(VotedPres2020_selection)) %>%
  mutate(VoteBiden = case_when(
    VotedPres2020_selection == "Biden" ~ 1,
    TRUE ~ 0
  ))
```

Let's first look at the main effects of income grouping and early voting behavior.

```
log_biden_main <- anes_des_ind %>%
  mutate(
    EarlyVote2020 = fct_relevel(EarlyVote2020, "No", after = 0)
  ) %>%
  svyglm(
    design = .,
    formula = VoteBiden ~ EarlyVote2020 + Income7,
    family = quasibinomial
  )
```

```
tidy(log_biden_main) %>%
  mutate(p.value = pretty_p_value(p.value)) %>%
  gt() %>%
  fmt_number()
```

TABLE 7.8 Logistic regression output for predicting voting for Biden given early voting behavior and income; main effects only, ANES 2020

term	estimate	std.error	statistic	p.value
(Intercept)	1.28	0.43	2.99	0.0047
EarlyVote2020Yes	0.44	0.34	1.29	0.2039
Income7$20k to < 40k	−1.06	0.49	−2.18	0.0352
Income7$40k to < 60k	−0.78	0.42	−1.86	0.0705
Income7$60k to < 80k	−1.24	0.70	−1.77	0.0842
Income7$80k to < 100k	−0.66	0.64	−1.02	0.3137
Income7$100k to < 125k	−1.02	0.54	−1.89	0.0662
Income7$125k or more	−1.25	0.44	−2.87	0.0065

This main effect model (see Table 7.8) indicates that people with incomes of $125,000 or more have a significant negative coefficient −1.25 (p-value is

0.0065). This indicates that people with incomes of $125,000 or more were less likely to vote for Biden in the 2020 election compared to people with incomes of $20,000 or less (reference level).

Although early voting behavior was not significant, there may be an interaction between income and early voting behavior. To determine this, we can create a model that includes the interaction effects:

```
log_biden_int <- anes_des_ind %>%
  mutate(
    EarlyVote2020 = fct_relevel(EarlyVote2020, "No", after = 0)
  ) %>%
  svyglm(
    design = .,
    formula = VoteBiden ~ (EarlyVote2020 + Income7)^2,
    family = quasibinomial
  )
```

```
tidy(log_biden_int) %>%
  mutate(p.value = pretty_p_value(p.value)) %>%
  gt() %>%
  fmt_number()
```

The results from the interaction model (see Table 7.9) show that one interaction between early voting behavior and income is significant. To better understand what this interaction means, we can plot the predicted probabilities with an interaction plot. Let's first obtain the predicted probabilities for each possible combination of variables using the augment() function.

```
log_biden_pred <- log_biden_int %>%
  augment(type.predict = "response") %>%
  mutate(
    .se.fit = sqrt(attr(.fitted, "var")),
    .fitted = as.numeric(.fitted)
  ) %>%
  select(VoteBiden, EarlyVote2020, Income7, .fitted, .se.fit)
```

The y-axis is the predicted probabilities, one of our x-variables is on the x-axis, and the other is represented by multiple lines. Figure 7.4 shows the interaction plot with early voting behavior on the x-axis and income represented by the lines.

TABLE 7.9 Logistic regression output for predicting voting for Biden given early voting behavior and income; with interaction, ANES 2020

term		estimate	std.error	statistic	p.value
(Intercept)		2.32	0.67	3.45	0.0015
EarlyVote2020Yes		−0.81	0.78	−1.03	0.3081
Income7$20k to < 40k		−2.33	0.87	−2.68	0.0113
Income7$40k to < 60k		−1.67	0.89	−1.87	0.0700
Income7$60k to < 80k		−2.05	1.05	−1.96	0.0580
Income7$80k to < 100k		−3.42	1.12	−3.06	0.0043
Income7$100k to < 125k		−2.33	1.07	−2.17	0.0368
Income7$125k or more		−2.09	0.92	−2.28	0.0289
EarlyVote2020Yes:	Income7$20k to < 40k	1.60	0.95	1.69	0.1006
EarlyVote2020Yes:	Income7$40k to < 60k	0.99	1.00	0.99	0.3289
EarlyVote2020Yes:	Income7$60k to < 80k	0.90	1.14	0.79	0.4373
EarlyVote2020Yes:	Income7$80k to < 100k	3.22	1.16	2.78	0.0087
EarlyVote2020Yes:	Income7$100k to < 125k	1.64	1.11	1.48	0.1492
EarlyVote2020Yes:	Income7$125k or more	1.00	1.14	0.88	0.3867

```
log_biden_pred %>%
  filter(VoteBiden == 1) %>%
  distinct() %>%
  arrange(EarlyVote2020, Income7) %>%
  ggplot(aes(
    x = EarlyVote2020,
    y = .fitted,
    group = Income7,
    color = Income7,
    linetype = Income7
  )) +
  geom_line(linewidth = 1.1) +
  scale_color_manual(values = colorRampPalette(book_colors)(7)) +
  ylab("Predicted Probability of Voting for Biden") +
  labs(
    x = "Voted Early",
    color = "Income",
```

```
    linetype = "Income"
) +
coord_cartesian(ylim = c(0, 1)) +
guides(fill = "none") +
theme_minimal()
```

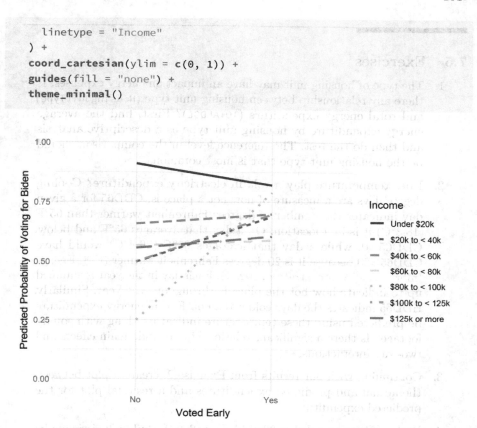

FIGURE 7.4 Interaction plot of early voting and income predicting the probability of voting for Biden

From Figure 7.4, we can see that people who have incomes in most groups (e.g., $40,000 to less than $60,000) have roughly the same probability of voting for Biden regardless of whether they voted early or not. However, those with income in the $100,000 to less than $125,000 group were more likely to vote for Biden if they voted early than if they did not vote early.

Interactions in models can be difficult to understand from the coefficients alone. Using these interaction plots can help others understand the nuances of the results.

7.5   Exercises

1. The type of housing unit may have an impact on energy expenses. Is there any relationship between housing unit type (HousingUnitType) and total energy expenditure (TOTALDOL)? First, find the average energy expenditure by housing unit type as a descriptive analysis and then do the test. The reference level in the comparison should be the housing unit type that is most common.

2. Does temperature play a role in electricity expenditure? Cooling degree days are a measure of how hot a place is. CDD65 for a given day indicates the number of degrees Fahrenheit warmer than 65°F (18.3°C) it is in a location. On a day that averages 65°F and below, CDD65=0, while a day that averages 85°F (29.4°C) would have CDD65=20 because it is 20 degrees Fahrenheit warmer (U.S. Energy Information Administration, 2023d). Each day in the year is summed up to indicate how hot the place is throughout the year. Similarly, HDD65 indicates the days colder than 65°F. Can energy expenditure be predicted using these temperature indicators along with square footage? Is there a significant relationship? Include main effects and two-way interactions.

3. Continuing with our results from Exercise 2, create a plot between the actual and predicted expenditures and a residual plot for the predicted expenditures.

4. Early voting expanded in 2020 (Sprunt, 2020). Build a logistic model predicting early voting in 2020 (EarlyVote2020) using age (Age), education (Education), and party identification (PartyID). Include two-way interactions.

5. Continuing from Exercise 4, predict the probability of early voting for two people. Both are 28 years old and have a graduate degree; however, one person is a strong Democrat, and the other is a strong Republican.

Part III

Reporting

Part 10

Reporting

8

Communication of results

Prerequisites

For this chapter, load the following packages:

```
library(tidyverse)
library(survey)
library(srvyr)
library(srvyrexploR)
library(gt)
library(gtsummary)
```

We are using data from ANES as described in Chapter 4. As a reminder, here is the code to create the design objects for each to use throughout this chapter. For ANES, we need to adjust the weight so it sums to the population instead of the sample (see the ANES documentation and Chapter 4 for more information).

```
targetpop <- 231592693

anes_adjwgt <- anes_2020 %>%
  mutate(Weight = Weight / sum(Weight) * targetpop)

anes_des <- anes_adjwgt %>%
  as_survey_design(
    weights = Weight,
    strata = Stratum,
    ids = VarUnit,
    nest = TRUE
  )
```

8.1   Introduction

After finishing the analysis and modeling, we proceed to the task of communicating the survey results. Our audience may range from seasoned researchers
familiar with our survey data to newcomers encountering the information for
the first time. We should aim to explain the methodology and analysis while
presenting findings in an accessible way, and it is our responsibility to report
information with care.

Before beginning any dissemination of results, consider questions such as:

- How are we presenting results? Examples include a website, print, or other
  media. Based on the medium, we might limit or enhance the use of graphical
  representation.
- What is the audience's familiarity with the study and/or data? Audiences
  can range from the general public to data experts. If we anticipate limited
  knowledge about the study, we should provide detailed descriptions (we
  discuss recommendations later in the chapter).
- What are we trying to communicate? It could be summary statistics, trends,
  patterns, or other insights. Tables may suit summary statistics, while plots
  are better at conveying trends and patterns.
- Is the audience accustomed to interpreting plots? If not, include explanatory
  text to guide them on how to interpret the plots effectively.
- What is the audience's statistical knowledge? If the audience does not have
  a strong statistics background, provide text on standard errors, confidence
  intervals, and other estimate types to enhance understanding.

8.2   Describing results through text

As analysts, we often emphasize the data, and communicating results can
sometimes be overlooked. To be effective communicators, we need to identify
the appropriate information to share with our audience. Chapters 2 and 3
provide insights into factors we need to consider during analysis, and they
remain relevant when presenting results to others.

8.2.1   Methodology

If we are using existing data, methodologically sound surveys provide documentation about how the survey was fielded, the questionnaires, and other
necessary information for analyses. For example, the survey's methodology re-
ports should include the population of interest, sampling procedures, response

rates, questionnaire documentation, weighting, and a general overview of disclosure statements. Many American organizations follow the American Association for Public Opinion Research's (AAPOR) Transparency Initiative[1]. The AAPOR Transparency Initiative requires organizations to include specific details in their methodology, making it clear how we can and should analyze and interpret the results. Being transparent about these methods is vital for the scientific rigor of the field.

The details provided in Chapter 2 about the survey process should be shared with the audience when presenting the results. When using publicly available data, like the examples in this book, we can often link to the methodology report in our final output. We should also provide high-level information for the audience to quickly grasp the context around the findings. For example, we can mention when and where the study was conducted, the population's age range, or other contextual details. This information helps the audience understand how generalizable the results are.

Providing this material is especially important when no methodology report is available for the analyzed data. For example, if we conducted a new survey for a specific purpose, we should document and present all the pertinent information during the analysis and reporting process. Adhering to the AAPOR Transparency Initiative guidelines is a reliable method to guarantee that all essential information is communicated to the audience.

8.2.2 Analysis

Along with the survey methodology and weight calculations, we should also share our approach to preparing, cleaning, and analyzing the data. For example, in Chapter 6, we compared education distributions from the ANES survey to the American Community Survey (ACS). To make the comparison, we had to collapse the education categories provided in the ANES data to match the ACS. The process for this particular example may seem straightforward (like combining bachelor's and graduate degrees into a single category), but there are multiple ways to deal with the data. Our choice is just one of many. We should document both the original ANES question and response options and the steps we took to match them with ACS data. This transparency helps clarify our analysis to our audience.

Missing data is another instance where we want to be unambiguous and upfront with our audience. In this book, numerous examples and exercises remove missing data, as this is often the easiest way to handle them. However, there are circumstances where missing data holds substantive importance, and excluding them could introduce bias (see Chapter 11). Being transparent about our handling of missing data is important to maintaining the integrity of our analysis and ensuring a comprehensive understanding of the results.

---

[1]https://aapor.org/standards-and-ethics/transparency-initiative

8.2.3   Results

While tables and graphs are commonly used to communicate results, there are instances where text can be more effective in sharing information. Narrative details, such as context around point estimates or model coefficients, can go a long way in improving our communication. We have several strategies to effectively convey the significance of the data to the audience through text.

First, we can highlight important data elements in a sentence using plain language. For example, if we were looking at election polling data conducted before an election, we could say:

As of [DATE], an estimated XX% of registered U.S. voters say they will vote for [CANDIDATE NAME] for president in the [YEAR] general election.

This sentence provides key pieces of information in a straightforward way:

1. [DATE]: Given that polling data are time-specific, providing the date of reference lets the audience know when these data were valid.
2. Registered U.S. voters: This tells the audience who we surveyed, letting them know the population of interest.
3. XX%: This part provides the estimated percentage of people voting for a specific candidate for a specific office.
4. [YEAR] general election: Adding this gives more context about the election type and year. The estimate would take on a different meaning if we changed it to a primary election instead of a general election.

We also included the word "estimated." When presenting aggregate survey results, we have errors around each estimate. We want to convey this uncertainty rather than talk in absolutes. Words like "estimated," "on average," or "around" can help communicate this uncertainty to the audience. Instead of saying "XX%," we can also say "XX% (+/- Y%)" to show the margin of error. Confidence intervals can also be incorporated into the text to assist readers.

Second, providing context and discussing the meaning behind a point estimate can help the audience glean some insight into why the data are important. For example, when comparing two values, it can be helpful to highlight if there are statistically significant differences and explain the impact and relevance of this information. This is where we should do our best to be mindful of biases and present the facts logically.

Keep in mind how we discuss these findings can greatly influence how the audience interprets them. If we include speculation, phrases like "the authors speculate" or "these findings may indicate," it relays the uncertainty around the notion while still lending a plausible solution. Additionally, we can present alternative viewpoints or competing discussion points to explain the uncertainty in the results.

8.3 Visualizing data

Although discussing key findings in the text is important, presenting large amounts of data in tables or visualizations is often more digestible for the audience. Effectively combining text, tables, and graphs can be powerful in communicating results. This section provides examples of using the {gt}, {gtsummary}, and {ggplot2} packages to enhance the dissemination of results (Iannone et al., 2024; Sjoberg et al., 2021; Wickham, 2016).

8.3.1 Tables

Tables are a great way to provide a large amount of data when individual data points need to be examined. However, it is important to present tables in a reader-friendly format. Numbers should align, rows and columns should be easy to follow, and the table size should not compromise readability. Using key visualization techniques, we can create tables that are informative and nice to look at. Many packages create easy-to-read tables (e.g., {kable} + {kable-Extra}, {gt}, {gtsummary}, {DT}, {formattable}, {flextable}, {reactable}). We appreciate the flexibility, ability to use pipes (e.g., %>%), and numerous extensions of the {gt} package. While we focus on {gt} here, we encourage learning about others, as they may have additional helpful features. Please note, at this time, {gtsummary} needs additional features to be widely used for survey analysis, particularly due to its lack of ability to work with replicate designs. We provide one example using {gtsummary} and hope it evolves into a more comprehensive tool over time.

Transitioning {srvyr} output to a {gt} table

Let's start by using some of the data we calculated earlier in this book. In Chapter 6, we looked at data on trust in government with the proportions calculated below:

```
trust_gov <- anes_des %>%
  drop_na(TrustGovernment) %>%
  group_by(TrustGovernment) %>%
```

```
summarize(trust_gov_p = survey_prop())

trust_gov
```

```
## # A tibble: 5 x 3
##    TrustGovernment      trust_gov_p trust_gov_p_se
##    <fct>                      <dbl>          <dbl>
## 1 Always                    0.0155        0.00204
## 2 Most of the time          0.132         0.00553
## 3 About half the time       0.309         0.00829
## 4 Some of the time          0.434         0.00855
## 5 Never                     0.110         0.00566
```

The default output generated by R may work for initial viewing inside our IDE or when creating basic output in an R Markdown or Quarto document. However, when presenting these results in other publications, such as the print version of this book or with other formal dissemination modes, modifying the display can improve our reader's experience.

Looking at the output from trust_gov, a couple of improvements stand out: (1) switching to percentages instead of proportions and (2) removing the variable names as column headers. The {gt} package is a good tool for implementing better labeling and creating publishable tables. Let's walk through some code as we make a few changes to improve the table's usefulness.

First, we initiate the formatted table with the gt() function on the trust_gov tibble previously created. Next, we use the argument rowname_col() to designate the TrustGovernment column as the label for each row (called the table "stub"). We apply the cols_label() function to create informative column labels instead of variable names and then the tab_spanner() function to add a label across multiple columns. In this case, we label all columns except the stub with "Trust in Government, 2020." We then format the proportions into percentages with the fmt_percent() function and reduce the number of decimals shown to one with decimals = 1. Finally, the tab_caption() function adds a table title for the HTML version of the book. We can use the caption for cross-referencing in R Markdown, Quarto, and bookdown, as well as adding it to the list of tables in the book. These changes are all seen in Table 8.1.

```
trust_gov_gt <- trust_gov %>%
  gt(rowname_col = "TrustGovernment") %>%
  cols_label(
    trust_gov_p = "%",
    trust_gov_p_se = "s.e. (%)"
  ) %>%
  tab_spanner(
```

```
  label = "Trust in Government, 2020",
  columns = c(trust_gov_p, trust_gov_p_se)
) %>%
fmt_percent(decimals = 1)
```

```
trust_gov_gt %>%
  tab_caption("Example of {gt} table with trust in government estimate")
```

TABLE 8.1 Example of {gt} table with trust in government estimate

	Trust in Government, 2020	
	%	s.e. (%)
Always	1.6%	0.2%
Most of the time	13.2%	0.6%
About half the time	30.9%	0.8%
Some of the time	43.4%	0.9%
Never	11.0%	0.6%

We can add a few more enhancements, such as a title (which is different from a caption[2]), a data source note, and a footnote with the question information, using the functions tab_header(), tab_source_note(), and tab_footnote(). If having the percentage sign in both the header and the cells seems redundant, we can opt for fmt_number() instead of fmt_percent() and scale the number by 100 with scale_by = 100. The resulting table is displayed in Table 8.2.

```
trust_gov_gt2 <- trust_gov_gt %>%
  tab_header("American voter's trust
              in the federal government, 2020") %>%
  tab_source_note(
    md("*Source*: American National Election Studies, 2020")
  ) %>%
  tab_footnote(
    "Question text: How often can you trust the federal government
    in Washington to do what is right?"
  ) %>%
```

---

[2]The tab_caption() function is intended for usage in R Markdown, Quarto, or bookdown to add cross-references across the document. The caption is placed within the table based on the output type. The tab_header() function adds a title or subtitle to a table in any context, including Shiny or GitHub-flavored Markdown, without cross-referencing. The header is placed within the table object itself.

```
fmt_number(
  scale_by = 100,
  decimals = 1
)
```

trust_gov_gt2

TABLE 8.2 Example of {gt} table with trust in government estimates and additional context

American voter's trust in the federal government, 2020

	Trust in Government, 2020	
	%	s.e. (%)
Always	1.6	0.2
Most of the time	13.2	0.6
About half the time	30.9	0.8
Some of the time	43.4	0.9
Never	11.0	0.6

Question text: How often can you trust the federal government in Washington to do what is right?
Source: American National Election Studies, 2020

Expanding tables using {gtsummary}

The {gtsummary} package simultaneously summarizes data and creates publication-ready tables. Initially designed for clinical trial data, it has been extended to include survey analysis in certain capacities. At this time, it is only compatible with survey objects using Taylor's Series Linearization and not replicate methods. While it offers a restricted set of summary statistics, the following are available for categorical variables:

- {n} frequency
- {N} denominator, or respondent population
- {p} proportion (stylized as a percentage by default)
- {p.std.error} standard error of the sample proportion

- {deff} design effect of the sample proportion
- {n_unweighted} unweighted frequency
- {N_unweighted} unweighted denominator
- {p_unweighted} unweighted formatted proportion (stylized as a percentage by default)

The following summary statistics are available for continuous variables:

- {median} median
- {mean} mean
- {mean.std.error} standard error of the sample mean
- {deff} design effect of the sample mean
- {sd} standard deviation
- {var} variance
- {min} minimum
- {max} maximum
- {p#} any integer percentile, where # is an integer from 0 to 100
- {sum} sum

In the following example, we build a table using {gtsummary}, similar to the table in the {gt} example. The main function we use is tbl_svysummary(). In this function, we include the variables we want to analyze in the include argument and define the statistics we want to display in the statistic argument. To specify the statistics, we apply the syntax from the {glue} package, where we enclose the variables we want to insert within curly brackets. We must specify the desired statistics using the names listed above. For example, to specify that we want the proportion followed by the standard error of the proportion in parentheses, we use {p} ({p.std.error}). Table 8.3 displays the resulting table.

```
anes_des_gtsum <- anes_des %>%
  tbl_svysummary(
    include = TrustGovernment,
    statistic = list(all_categorical() ~ "{p} ({p.std.error})")
  )


anes_des_gtsum
```

TABLE 8.3 Example of {gtsummary} table with trust in government estimates

Characteristic	N = 231,034,125[1]
PRE: How often trust government in Washington to do what is right [revised]	
Always	1.6 (0.00)
Most of the time	13 (0.01)
About half the time	31 (0.01)
Some of the time	43 (0.01)
Never	11 (0.01)
Unknown	673,773

[1]% (SE(%))

The default table (shown in Table 8.3) includes the weighted number of missing (or Unknown) records. The standard error is reported as a proportion, while the proportion is styled as a percentage. In the next step, we remove the Unknown category by setting the missing argument to "no" and format the standard error as a percentage using the `digits` argument. To improve the table for publication, we provide a more polished label for the "TrustGovernment" variable using the `label` argument. The resulting table is displayed in Table 8.4.

```
anes_des_gtsum2 <- anes_des %>%
  tbl_svysummary(
    include = TrustGovernment,
    statistic = list(all_categorical() ~ "{p} ({p.std.error})"),
    missing = "no",
    digits = list(TrustGovernment ~ style_percent),
    label = list(TrustGovernment ~ "Trust in Government, 2020")
  )


anes_des_gtsum2
```

TABLE 8.4 Example of {gtsummary} table with trust in government estimates with labeling and digits options

Characteristic	N = 231,034,125[1]
Trust in Government, 2020	
Always	1.6 (0.2)
Most of the time	13 (0.6)
About half the time	31 (0.8)
Some of the time	43 (0.9)
Never	11 (0.6)

[1]% (SE(%))

Table 8.4 is closer to our ideal output, but we still want to make a few changes. To exclude the term "Characteristic" and the estimated population size (N), we can modify the header using the modify_header() function to update the label. Further adjustments can be made based on personal preferences, organizational guidelines, or other style guides. If we prefer having the standard error in the header, similar to the {gt} table, instead of in the footnote (the {gtsummary} default), we can make these changes by specifying stat_0 in the modify_header() function. Additionally, using modify_footnote() with update = everything() ~ NA removes the standard error from the footnote. After transforming the object into a {gt} table using as_gt(), we can add footnotes and a title using the same methods explained in the previous section. This updated table is displayed in Table 8.5.

```
anes_des_gtsum3 <- anes_des %>%
  tbl_svysummary(
    include = TrustGovernment,
    statistic = list(all_categorical() ~ "{p} ({p.std.error})"),
    missing = "no",
    digits = list(TrustGovernment ~ style_percent),
    label = list(TrustGovernment ~ "Trust in Government, 2020")
  ) %>%
  modify_footnote(update = everything() ~ NA) %>%
  modify_header(
    label = " ",
    stat_0 = "% (s.e.)"
  ) %>%
  as_gt() %>%
  tab_header("American voter's trust
             in the federal government, 2020") %>%
  tab_source_note(
    md("*Source*: American National Election Studies, 2020")
```

```
) %>%
tab_footnote(
  "Question text: How often can you trust the federal government
in Washington to do what is right?"
)
```

```
anes_des_gtsum3
```

TABLE 8.5 Example of {gtsummary} table with trust in government estimates with more labeling options and context

American voter's trust in the federal government, 2020

	% (s.e.)
Trust in Government, 2020	
Always	1.6 (0.2)
Most of the time	13 (0.6)
About half the time	31 (0.8)
Some of the time	43 (0.9)
Never	11 (0.6)

Question text: How often can you trust the federal government in Washington to do what is right?
Source: American National Election Studies, 2020

We can also include summaries of more than one variable in the table. These variables can be either categorical or continuous. In the following code and Table 8.6, we add the mean age by updating the `include`, `statistic`, and `digits` arguments.

```
anes_des_gtsum4 <- anes_des %>%
  tbl_svysummary(
    include = c(TrustGovernment, Age),
    statistic = list(
      all_categorical() ~ "{p} ({p.std.error})",
      all_continuous() ~ "{mean} ({mean.std.error})"
    ),
    missing = "no",
    digits = list(TrustGovernment ~ style_percent,
                  Age ~ c(1, 2)),
```

```
    label = list(TrustGovernment ~ "Trust in Government, 2020")
) %>%
modify_footnote(update = everything() ~ NA) %>%
modify_header(label = " ",
              stat_0 = "% (s.e.)") %>%
as_gt() %>%
tab_header(
  "American voter's trust in the federal government, 2020") %>%
tab_source_note(
  md("*Source*: American National Election Studies, 2020")
) %>%
tab_footnote(
  "Question text: How often can you trust the federal government
  in Washington to do what is right?"
) %>%
tab_caption("Example of {gtsummary} table with trust in government
             estimates and average age")
```

anes_des_gtsum4

TABLE 8.6 Example of {gtsummary} table with trust in government estimates and average age

American voter's trust in the federal government, 2020

	% (s.e.)
Trust in Government, 2020	
Always	1.6 (0.2)
Most of the time	13 (0.6)
About half the time	31 (0.8)
Some of the time	43 (0.9)
Never	11 (0.6)
PRE: SUMMARY: Respondent age	47.3 (0.36)

Question text: How often can you trust the federal government in Washington to do what is right?
Source: American National Election Studies, 2020

With {gtsummary}, we can also calculate statistics by different groups. Let's modify the previous example (displayed in Table 8.6) to analyze data on

whether a respondent voted for president in 2020. We update the by argument
and refine the header. The resulting table is displayed in Table 8.7.

```
anes_des_gtsum5 <- anes_des %>%
  drop_na(VotedPres2020) %>%
  tbl_svysummary(
    include = TrustGovernment,
    statistic = list(all_categorical() ~ "{p} ({p.std.error})"),
    missing = "no",
    digits = list(TrustGovernment ~ style_percent),
    label = list(TrustGovernment ~ "Trust in Government, 2020"),
    by = VotedPres2020
  ) %>%
  modify_footnote(update = everything() ~ NA) %>%
  modify_header(
    label = " ",
    stat_1 = "Voted",
    stat_2 = "Didn't vote"
  ) %>%
  modify_spanning_header(all_stat_cols() ~ "% (s.e.)") %>%
  as_gt() %>%
  tab_header(
    "American voter's trust
                in the federal government by whether they voted
                in the 2020 presidential election"
  ) %>%
  tab_source_note(
    md("*Source*: American National Election Studies, 2020")
  ) %>%
  tab_footnote(
    "Question text: How often can you trust the federal government
    in Washington to do what is right?"
  )

anes_des_gtsum5
```

TABLE 8.7 Example of {gtsummary} table with trust in government estimates by voting status

American voter's trust in the federal government by whether they voted in the 2020 presidential election

	% (s.e.)	
	Voted	Didn't vote
Trust in Government, 2020		
Always	1.1 (0.2)	0.9 (0.9)
Most of the time	13 (0.6)	19 (5.3)
About half the time	32 (0.8)	30 (8.6)
Some of the time	45 (0.8)	45 (8.2)
Never	9.1 (0.7)	5.2 (2.2)

Question text: How often can you trust the federal government in Washington to do what is right?
Source: American National Election Studies, 2020

8.3.2 Charts and plots

Survey analysis can yield an abundance of printed summary statistics and models. Even with the most careful analysis, interpreting the results can be overwhelming. This is where charts and plots play a key role in our work. By transforming complex data into a visual representation, we can recognize patterns, relationships, and trends with greater ease.

R has numerous packages for creating compelling and insightful charts. In this section, we focus on {ggplot2}, a member of the {tidyverse} collection of packages. Known for its power and flexibility, {ggplot2} is an invaluable tool for creating a wide range of data visualizations (Wickham, 2016).

The {ggplot2} package follows the "grammar of graphics," a framework that incrementally adds layers of chart components. This approach allows us to customize visual elements such as scales, colors, labels, and annotations to enhance the clarity of our results. After creating the survey design object, we can modify it to include additional outcomes and calculate estimates for our desired data points. Below, we create a binary variable TrustGovernmentUsually, which is TRUE when TrustGovernment is "Always" or "Most of the time" and FALSE otherwise. Then, we calculate the percentage of people who usually trust the government based on their vote in the 2020 presidential election (VotedPres2020_selection). We remove the cases where people did not vote or did not indicate their choice.

```
anes_des_der <- anes_des %>%
  mutate(TrustGovernmentUsually = case_when(
    is.na(TrustGovernment) ~ NA,
    TRUE ~ TrustGovernment %in% c("Always", "Most of the time")
  )) %>%
  drop_na(VotedPres2020_selection) %>%
  group_by(VotedPres2020_selection) %>%
  summarize(
    pct_trust = survey_mean(
      TrustGovernmentUsually,
      na.rm = TRUE,
      proportion = TRUE,
      vartype = "ci"
    ),
    .groups = "drop"
  )

anes_des_der
```

```
## # A tibble: 3 x 4
##   VotedPres2020_selection pct_trust pct_trust_low pct_trust_upp
##   <fct>                       <dbl>         <dbl>         <dbl>
## 1 Biden                       0.123         0.109         0.140
## 2 Trump                       0.178         0.161         0.198
## 3 Other                      0.0681        0.0290         0.152
```

Now, we can begin creating our chart with {ggplot2}. First, we set up our
plot with ggplot(). Next, we define the data points to be displayed using
aesthetics, or aes. Aesthetics represent the visual properties of the objects in
the plot. In the following example, we create a bar chart of the percentage
of people who usually trust the government by who they voted for in the
2020 election. To do this, we want to have who they voted for on the x-axis
(VotedPres2020_selection) and the percent they usually trust the government
on the y-axis (pct_trust). We specify these variables in ggplot() and then
indicate we want a bar chart with geom_bar(). The resulting plot is displayed
in Figure 8.1.

```
p <- anes_des_der %>%
  ggplot(aes(
    x = VotedPres2020_selection,
    y = pct_trust
  )) +
  geom_bar(stat = "identity")

p
```

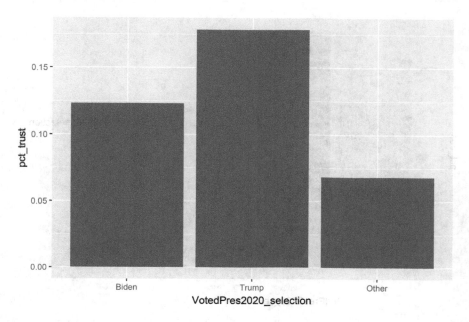

FIGURE 8.1 Bar chart of trust in government, by chosen 2020 presidential candidate

This is a great starting point: it appears that a higher percentage of people state they usually trust the government among those who voted for Trump compared to those who voted for Biden or other candidates. Now, what if we want to introduce color to better differentiate the three groups? We can add `fill` under `aesthetics`, indicating that we want to use distinct colors for each value of `VotedPres2020_selection`. In this instance, Biden and Trump are displayed in different colors in Figure 8.2.

```
pcolor <- anes_des_der %>%
  ggplot(aes(
    x = VotedPres2020_selection,
    y = pct_trust,
    fill = VotedPres2020_selection
  )) +
  geom_bar(stat = "identity")

pcolor
```

Let's say we wanted to follow proper statistical analysis practice and incorporate variability in our plot. We can add another geom, `geom_errorbar()`, to display

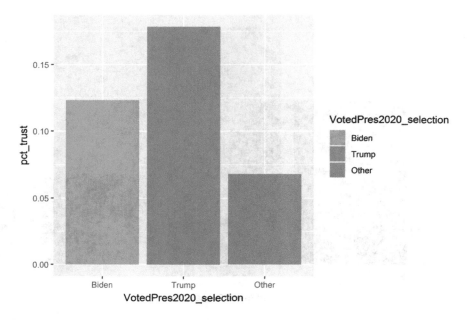

FIGURE 8.2 Bar chart of trust in government by chosen 2020 presidential candidate, with colors

the confidence intervals on top of our existing `geom_bar()` layer. We can add the layer using a plus sign (+). The resulting graph is displayed in Figure 8.3.

```
pcol_error <- anes_des_der %>%
  ggplot(aes(
    x = VotedPres2020_selection,
    y = pct_trust,
    fill = VotedPres2020_selection
  )) +
  geom_bar(stat = "identity") +
  geom_errorbar(
    aes(
      ymin = pct_trust_low,
      ymax = pct_trust_upp
    ),
    width = .2
  )

pcol_error
```

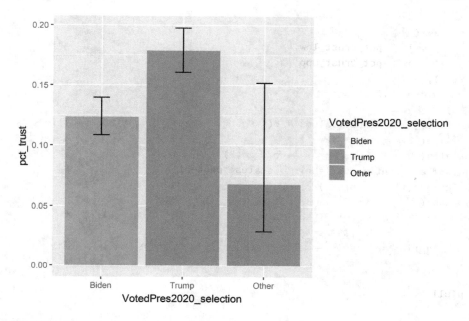

FIGURE 8.3 Bar chart of trust in government by chosen 2020 presidential candidate, with colors and error bars

We can continue adding to our plot until we achieve our desired look. For example, since the color legend does not contribute meaningful information, we can eliminate it with guides(fill = "none"). We can also specify colors for fill using scale_fill_manual(). Inside this function, we provide a vector of values corresponding to the colors in our plot. These values are hexadecimal (hex) color codes, denoted by a leading pound sign # followed by six letters or numbers. The hex code #0b3954 used below is dark blue. There are many tools online that help pick hex codes, such as htmlcolorcodes.com. Additionally, Figure 8.4 incorporates better labels for the x and y axes (xlab(), ylab()), a title (labs(title=)), and a footnote with the data source (labs(caption=)).

```
pfull <-
  anes_des_der %>%
  ggplot(aes(
    x = VotedPres2020_selection,
    y = pct_trust,
    fill = VotedPres2020_selection
)) +
  geom_bar(stat = "identity") +
  geom_errorbar(
```

```
aes(
  ymin = pct_trust_low,
  ymax = pct_trust_upp
),
width = .2
) +
scale_fill_manual(values = c("#0b3954", "#bfd7ea", "#8d6b94")) +
xlab("Election choice (2020)") +
ylab("Usually trust the government") +
scale_y_continuous(labels = scales::percent) +
guides(fill = "none") +
labs(
  title = "Percent of voters who usually trust the government
    by chosen 2020 presidential candidate",
  caption = "Source: American National Election Studies, 2020"
)

pfull
```

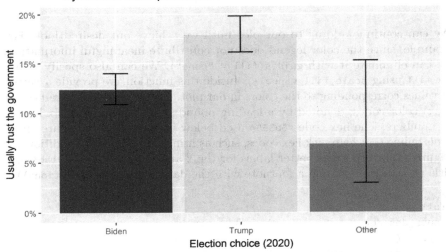

FIGURE 8.4 Bar chart of trust in government by chosen 2020 presidential candidate with colors, labels, error bars, and title

What we have explored in this section are just the foundational aspects of {ggplot2}, and the capabilities of this package extend far beyond what we have covered. Advanced features such as annotation, faceting, and theming allow for more sophisticated and customized visualizations. The {ggplot2} book by Wickham (2016) is a comprehensive guide to learning more about this powerful tool.

9

Reproducible research

9.1 Introduction

Reproducing results is an important aspect of any research. First, reproducibility serves as a form of quality assurance. If we pass an analysis project to another person, they should be able to run the entire project from start to finish and obtain the same results. They can critically assess the methodology and code while detecting potential errors. Another goal of reproducibility is enabling the verification of our analysis. When someone else is able to check our results, it ensures the integrity of the analyses by determining that the conclusions are not dependent on a particular person running the code or workflow on a particular day or in a particular environment.

Not only is reproducibility a key component in ethical and accurate research, but it is also a requirement for many scientific journals. For example, the *Journal of Survey Statistics and Methodology* (JSSAM) and *Public Opinion Quarterly* (POQ) require authors to make code, data, and methodology transparent and accessible to other researchers who wish to verify or build on existing work.

Reproducible research requires that the key components of analysis are available, discoverable, documented, and shared with others. The four main components that we should consider are:

- Code: source code used for data cleaning, analysis, modeling, and reporting
- Data: raw data used in the workflow, or if data are sensitive or proprietary, as much data as possible that would allow others to run our workflow or provide details on how to access the data (e.g., access to a restricted use file (RUF))
- Environment: environment of the project, including the R version, packages, operating system, and other dependencies used in the analysis
- Methodology: survey and analysis methodology, including rationale behind sample, questionnaire and analysis decisions, interpretations, and assumptions

In Chapter 8, we briefly mention how each of these is important to include in the methodology report and when communicating the findings of a study. However, to be transparent and effective analysts, we need to ensure we not only discuss these through text but also provide files and additional information when requested. Often, when starting a project, we may be eager to jump into the data and make decisions as we go without full documentation. This can be challenging if we need to go back and make changes or understand even what we did a few months ago. It benefits other analysts and potentially our future selves to document everything from the start. The good news is that many tools, practices, and project management techniques make survey analysis projects easy to reproduce. For best results, we should decide which techniques and tools to use before starting a project (or very early on).

This chapter covers some of our suggestions for tools and techniques we can use in projects. This list is not comprehensive but aims to provide a starting point for those looking to create a reproducible workflow.

9.2   Project-based workflows

We recommend a project-based workflow for analysis projects as described by Wickham et al. (2023c). A project-based workflow maintains a "source of truth" for our analyses. It helps with file system discipline by putting everything related to a project in a designated folder. Since all associated files are in a single location, they are easy to find and organize. When we reopen the project, we can recreate the environment in which we originally ran the code to reproduce our results.

The RStudio IDE has built-in support for projects. When we create a project in RStudio, it creates an .Rproj file that stores settings specific to that project. Once we have created a project, we can create folders that help us organize our workflow. For example, a project directory could look like this:

```
| anes_analysis/
  | anes_analysis.Rproj
  | README.md
  | codebooks
    | codebook2020.pdf
    | codebook2016.pdf
  | rawdata
    | anes2020_raw.csv
    | anes2016_raw.csv
  | scripts
    | data-prep.R
  | data
    | anes2020_clean.csv
    | anes2016_clean.csv
```

```
| report
  | anes_report.Rmd
  | anes_report.html
  | anes_report.pdf
```

In a project-based workflow, all paths are relative and, by default, relative to the folder the .Rproj file is located in. By using relative paths, others can open and run our files even if their directory configuration differs from ours (e.g., Mac and Windows users have different directory path structures). The {here} package enables easy file referencing, and we can start by using the here::here() function to build the path for loading or saving data (Müller, 2020). Below, we ask R to read the CSV file anes_2020.csv in the project directory's data folder:

```
anes <-
  read_csv(here::here("data", "anes2020_clean.csv"))
```

The combination of projects and the {here} package keep all associated files organized. This workflow makes it more likely that our analyses can be reproduced by us or our colleagues.

9.3 Functions and packages

We may find that we are repeating ourselves in our script, and the chance of errors increases whenever we copy and paste our code. By creating a function, we can create a consistent set of commands that reduce the likelihood of mistakes. Functions also organize our code, improve the code readability, and allow others to execute the same commands. For example, in Chapter 13, we create a function to run sequences of rename(), filter(), group_by(), and summarize statements across different variables. Creating functions helps us avoid overlooking necessary steps.

A package is made up of a collection of functions. If we find ourselves sharing functions with others to replicate the same series of commands in a separate project, creating a package can be a useful tool for sharing the code along with data and documentation.

9.4　Version control with Git

Often, a survey analysis project produces a lot of code. Keeping track of the latest version can become challenging, as files evolve throughout a project. If a team of analysts is working on the same script, someone may use an outdated version, resulting in incorrect results or redundant work.

Version control systems like Git can help alleviate these pains. Git is a system that tracks changes in files. We can use Git to follow code evaluation and manage asynchronous work. With Git, it is easy to see any changes made in a script, revert changes, and resolve differences between code versions (called conflicts).

Services such as GitHub or GitLab provide hosting and sharing of files as well as version control with Git. For example, we can visit the GitHub repository for this book[1] and see the files that build the book, when they were committed to the repository, and the history of modifications over time.

In addition to code scripts, platforms like GitHub can store data and documentation. They provide a way to maintain a history of data modifications through versioning and timestamps. By saving the data and documentation alongside the code, it becomes easier for others to refer to and access everything they need in one place.

Using version control in analysis projects makes collaboration and maintenance more manageable. To connect Git with R, we recommend referencing the book Happy Git and GitHub for the useR[2] (Bryan, 2023).

9.5　Package management with {renv}

Ensuring reproducibility involves not only using version control of code but also managing the versions of packages. If two people run the same code but use different package versions, the results might differ because of changes to those packages. For example, this book currently uses a version of the {srvyr} package from GitHub and not from CRAN. This is because the version of {srvyr} on CRAN has some bugs (errors) that result in incorrect calculations. The version on GitHub has corrected these errors, so we have asked readers to install the GitHub version to obtain the same results.

---

[1]https://github.com/tidy-survey-r/tidy-survey-book
[2]https://happygitwithr.com/

One way to handle different package versions is with the {renv} package. This package allows researchers to set the versions for each used package and manage package dependencies. Specifically, {renv} creates isolated, project-specific environments that record the packages and their versions used in the code. When initiated by a new user, {renv} checks whether the installed packages are consistent with the recorded version for the project. If not, it installs the appropriate versions so that others can replicate the project's environment to rerun the code and obtain consistent results (Ushey and Wickham, 2024).

9.6 R environments with Docker

Just as different versions of packages can introduce discrepancies or compatibility issues, the version of R can also prevent reproducibility. Tools such as Docker can help with this potential issue by creating isolated environments that define the version of R being used, along with other dependencies and configurations. The entire environment is bundled in a container. The container, defined by a Dockerfile, can be shared so that anybody, regardless of their local setup, can run the R code in the same environment.

9.7 Workflow management with {targets}

With complex studies involving multiple code files and dependencies, it is important to ensure each step is executed in the intended sequence. We can do this manually, e.g., by numbering files to indicate the order or providing detailed documentation on the order. Alternatively, we can automate the process so the code flows sequentially. Making sure that the code runs in the correct order helps ensure that the research is reproducible. Anyone should be able to pick up the set of scripts and get the same results by following the workflow.

The {targets} package is an increasingly popular workflow manager that documents, automates, and executes complex data workflows with multiple steps and dependencies. With this package, we first define the order of execution for our code, and then it consistently executes the code in that order each time it is run. One beneficial feature of {targets} is that if code changes later in the workflow, only the affected code and its downstream targets (i.e., the subsequent code files) are re-executed when we change a script. The {targets} package also provides interactive progress monitoring and reporting, allowing us to track the status and progress of our analysis pipeline (Landau, 2021).

9.8   Documentation with Quarto and R Markdown

Tools like Quarto and R Markdown aid in reproducibility by creating documents that weave together code, text, and results. We can present analysis results alongside the report's narrative, so there's no need to copy and paste code output into the final documentation. By eliminating manual steps, we can reduce the chances of errors in the final output.

Quarto and R Markdown documents also allow users to re-execute the underlying code when needed. Another analyst can see the steps we took, follow the scripts, and recreate the report. We can include details about our work in one place thanks to the combination of text and code, making our work transparent and easier to verify (Allaire et al., 2024; Xie et al., 2020).

9.8.1   Parameterization

Another useful feature of Quarto and R Markdown is the ability to reduce repetitive code by parameterizing the files. Parameters can control various aspects of the analysis, such as dates, geography, or other analysis variables. We can define and modify these parameters to explore different scenarios or inputs. For example, suppose we start by creating a document that provides survey analysis results for North Carolina but then later decide we want to look at another state. In that case, we can define a `state` parameter and rerun the same analysis for a state like Washington without having to edit the code throughout the document.

Parameters can be defined in the header or code chunks of our Quarto or R Markdown documents and easily modified and documented. By manually editing code throughout the script, we reduce errors that may occur and offer a flexible way for others to replicate the analysis and explore variations.

9.9   Other tips for reproducibility

9.9.1   Random number seeds

Some tasks in survey analysis require randomness, such as imputation, model training, or creating random samples. By default, the random numbers generated by R change each time we rerun the code, making it difficult to reproduce the same results. By "setting the seed," we can control the randomness and ensure that the random numbers remain consistent whenever we rerun the code. Others can use the same seed value to reproduce our random numbers and achieve the same results.

In R, we can use the `set.seed()` function to control the randomness in our code. We set a seed value by providing an integer in the function argument. The following code chunk sets a seed using `999`, then runs a random number function (`runif()`) to get five random numbers from a uniform distribution.

```
set.seed(999)
runif(5)
```

```
## [1] 0.38907 0.58306 0.09467 0.85263 0.78675
```

Since the seed is set to `999`, running `runif(5)` multiple times always produces the same output. The choice of the seed number is up to the analyst. For example, this could be the date (`20240102`) or time of day (`1056`) when the analysis was first conducted, a phone number (`8675309`), or the first few numbers that come to mind (`369`). As long as the seed is set for a given analysis, the actual number is up to the analyst to decide. It is important to note that `set.seed()` should be used before random number generation. Run it once per program, and the seed is applied to the entire script. We recommend setting the seed at the beginning of a script, where libraries are loaded.

9.9.2 Descriptive names and labels

Using descriptive variable names or labeling data can also assist with reproducible research. For example, in the ANES data, the variable names in the raw data all start with `V20` and are a string of numbers. To make things easier to reproduce in this book, we opted to change the variable names to be more descriptive of what they contained (e.g., `Age`). This can also be done with the data values themselves. One way to accomplish this is by creating factors for categorical data, which can ensure that we know that a value of `1` really means `Female`, for example. There are other ways of handling this, such as attaching labels to the data instead of recoding variables to be descriptive (see Chapter 11). As with random number seeds, the exact method is up to the analyst, but providing this information can help ensure our research is reproducible.

9.10 Additional resources

We can promote accuracy and verification of results by making our analysis reproducible. There are various tools and guides available to help achieve reproducibility in analysis work, a few of which were described in this chapter. Here are additional resources to explore:

- R for Data Science chapter on project-based workflows[3]
- Building reproducible analytical pipelines with R[4]
- Posit Solutions Site page on reproducible environments[5]

---

[3]https://r4ds.hadley.nz/workflow-scripts.html#projects
[4]https://raps-with-r.dev/
[5]https://solutions.posit.co/envs-pkgs/environments/

Part IV

Real-life data

10

Sample designs and replicate weights

10.1 Introduction

The primary reason for using packages like {survey} and {srvyr} is to account for the sampling design or replicate weights into point and uncertainty estimates (Freedman Ellis and Schneider, 2024; Lumley, 2010). By incorporating the sampling design or replicate weights, these estimates are appropriately calculated.

In this chapter, we introduce common sampling designs and common types of replicate weights, the mathematical methods for calculating estimates and standard errors for a given sampling design, and the R syntax to specify the sampling design or replicate weights. While we show the math behind the estimates, the functions in these packages handle the calculation. To deeply understand the math and the derivation, refer to Penn State (2019), Särndal et al. (2003), Wolter (2007), or Fuller (2011) (these are listed in order of increasing statistical rigorousness).

The general process for estimation in the {srvyr} package is to:

1. Create a `tbl_svy` object (a survey object) using: `as_survey_design()` or `as_survey_rep()`

2. Subset data (if needed) using `filter()` (subpopulations)

3. Specify domains of analysis using `group_by()`

4. Within `summarize()`, specify variables to calculate, including means, totals, proportions, quantiles, and more

This chapter includes details on the first step: creating the survey object. Once this survey object is created, it can be used in the other steps (detailed in Chapters 5 through 7) to account for the complex survey design.

10.2   Common sampling designs

A sampling design is the method used to draw a sample. Both logistical and statistical elements are considered when developing a sampling design. When specifying a sampling design in R, we specify the levels of sampling along with the weights. The weight for each record is constructed so that the particular record represents that many units in the population. For example, in a survey of 6th-grade students in the United States, the weight associated with each responding student reflects how many 6th-grade students across the country that record represents. Generally, the weights represent the inverse of the probability of selection, such that the sum of the weights corresponds to the total population size, although some studies may have the sum of the weights equal to the number of respondent records.

Some common terminology across the designs are:

- sample size, generally denoted as n, is the number of units selected to be sampled
- population size, generally denoted as N, is the number of units in the population of interest
- sampling frame, the list of units from which the sample is drawn (see Chapter 2 for more information)

10.2.1   Simple random sample without replacement

The simple random sample (SRS) without replacement is a sampling design in which a fixed sample size is selected from a sampling frame, and every possible subsample has an equal probability of selection. Without replacement refers to the fact that once a sampling unit has been selected, it is removed from the sample frame and cannot be selected again.

- Requirements: The sampling frame must include the entire population.
- Advantages: SRS requires no information about the units apart from contact information.
- Disadvantages: The sampling frame may not be available for the entire population.
- Example: Randomly select students in a university from a roster provided by the registrar's office.

The math

The estimate for the population mean of variable y is:

$$\bar{y} = \frac{1}{n} \sum_{i=1}^{n} y_i$$

where $\bar{y}$ represents the sample mean, n is the total number of respondents (or observations), and y_i is each individual value of y.

The estimate of the standard error of the mean is:

$$se(\bar{y}) = \sqrt{\frac{s^2}{n} \left(1 - \frac{n}{N}\right)}$$

where

$$s^2 = \frac{1}{n-1} \sum_{i=1}^{n} (y_i - \bar{y})^2.$$

and N is the population size. This standard error estimate might look very similar to equations in other statistical applications except for the part on the right side of the equation: $1 - \frac{n}{N}$. This is called the finite population correction (FPC) factor. If the size of the frame, N, is very large in comparison to the sample, the FPC is negligible, so it is often ignored. A common guideline is if the sample is less than 10% of the population, the FPC is negligible.

To estimate proportions, we define x_i as the indicator if the outcome is observed. That is, $x_i = 1$ if the outcome is observed, and $x_i = 0$ if the outcome is not observed for respondent i. Then the estimated proportion from an SRS design is:

$$\hat{p} = \frac{1}{n} \sum_{i=1}^{n} x_i$$

and the estimated standard error of the proportion is:

$$se(\hat{p}) = \sqrt{\frac{\hat{p}(1-\hat{p})}{n-1}\left(1 - \frac{n}{N}\right)}$$

The syntax

If a sample was drawn through SRS and had no nonresponse or other weighting adjustments, we specify this design in R as:

```
srs1_des <- dat %>%
  as_survey_design(fpc = fpcvar)
```

where `dat` is a tibble or data.frame with the survey data, and `fpcvar` is a variable in the data indicating the sampling frame's size (this variable has the same value for all cases in an SRS design). If the frame is very large, sometimes the frame size is not provided. In that case, the FPC is not needed, and we specify the design as:

```
srs2_des <- dat %>%
  as_survey_design()
```

If some post-survey adjustments were implemented and the weights are not all equal, we specify the design as:

```
srs3_des <- dat %>%
  as_survey_design(weights = wtvar,
                   fpc = fpcvar)
```

where `wtvar` is a variable in the data indicating the weight for each case. Again, the FPC can be omitted if it is unnecessary because the frame is large compared to the sample size.

Example

The {survey} package in R provides some example datasets that we use throughout this chapter. One of the example datasets we use is from the Academic Performance Index Program (APIP). The APIP program administered by the California Department of Education, and the {survey} package includes a population file (sample frame) of all schools with at least 100 students and several different samples pulled from that data using different sampling methods. For this first example, we use the `apisrs` dataset, which contains an SRS of 200 schools. For printing purposes, we create a new dataset called `apisrs_slim`, which sorts the data by the school district and school ID and subsets the data to only a few columns. The SRS sample data are illustrated below:

```
apisrs_slim <-
  apisrs %>%
  as_tibble() %>%
  arrange(dnum, snum) %>%
  select(cds, dnum, snum, dname, sname, fpc, pw)

apisrs_slim
```

```
## # A tibble: 200 x 7
##    cds            dnum  snum dname                sname  fpc    pw
##    <chr>         <int> <dbl> <chr>                <chr> <dbl> <dbl>
##  1 19642126061220    1  1121 ABC Unified          Hask~  6194  31.0
##  2 19642126066716    1  1124 ABC Unified          Stow~  6194  31.0
##  3 36675876035174    5  3895 Adelanto Elementary  Adel~  6194  31.0
##  4 33669776031512   19  3347 Alvord Unified       Arla~  6194  31.0
##  5 33669776031595   19  3352 Alvord Unified       Well~  6194  31.0
##  6 31667876031033   39  3271 Auburn Union Elementa~ Cain~ 6194  31.0
##  7 19642876011407   42  1169 Baldwin Park Unified Dean~  6194  31.0
##  8 19642876011464   42  1175 Baldwin Park Unified Heat~  6194  31.0
##  9 19642956011589   48  1187 Bassett Unified      Erwi~  6194  31.0
## 10 41688586043392   49  4948 Bayshore Elementary  Bays~  6194  31.0
## # i 190 more rows
```

Table 10.1 provides details on all the variables in this dataset.

TABLE 10.1 Overview of Variables in APIP Data

Variable Name	Description
cds	Unique identifier for each school
dnum	School district identifier within county
snum	School identifier within district
dname	District Name
sname	School Name
fpc	Finite population correction factor
pw	Weight

To create the `tbl_survey` object for the SRS data, we specify the design as:

```
apisrs_des <- apisrs_slim %>%
  as_survey_design(
    weights = pw,
    fpc = fpc
  )

apisrs_des
```

```
## Independent Sampling design
## Called via srvyr
## Sampling variables:
##    - ids: `1`
##    - fpc: fpc
##    - weights: pw
## Data variables:
##    - cds (chr), dnum (int), snum (dbl), dname (chr), sname (chr), fpc
##      (dbl), pw (dbl)
```

In the printed design object, the design is described as an "Independent Sampling design," which is another term for SRS. The ids are specified as 1, which means there is no clustering (a topic described in Section 10.2.4), the FPC variable is indicated, and the weights are indicated. We can also look at the summary of the design object (`summary()`) and see the distribution of the probabilities (inverse of the weights) along with the population size and a list of the variables in the dataset.

```
summary(apisrs_des)
```

```
## Independent Sampling design
## Called via srvyr
## Probabilities:
##    Min. 1st Qu.  Median    Mean 3rd Qu.    Max.
##  0.0323  0.0323  0.0323  0.0323  0.0323  0.0323
## Population size (PSUs): 6194
## Data variables:
## [1] "cds"    "dnum"   "snum"   "dname"  "sname"  "fpc"    "pw"
```

10.2.2   Simple random sample with replacement

Similar to the SRS design, the simple random sample with replacement (SRSWR) design randomly selects the sample from the entire sampling frame. However, while SRS removes sampled units before selecting again, the SRSWR instead replaces each sampled unit before drawing again, so units can be selected more than once.

- Requirements: The sampling frame must include the entire population.
- Advantages: SRSWR requires no information about the units apart from contact information.
- Disadvantages:
  - The sampling frame may not be available for the entire population.
  - Units can be selected more than once, resulting in a smaller realized sample size because receiving duplicate information from a single respondent does not provide additional information.
  - For small populations, SRSWR has larger standard errors than SRS designs.
- Example: A professor puts all students' names on paper slips and selects them randomly to ask students questions, but the professor replaces the paper after calling on the student so they can be selected again at any time.

In general for surveys, using an SRS design (without replacement) is preferred as we do not want respondents to answer a survey more than once.

The math

The estimate for the population mean of variable y is:

$$\bar{y} = \frac{1}{n} \sum_{i=1}^{n} y_i$$

and the estimate of the standard error of mean is:

$$se(\bar{y}) = \sqrt{\frac{s^2}{n}}$$

where

$$s^2 = \frac{1}{n-1} \sum_{i=1}^{n} (y_i - \bar{y})^2 .$$

To calculate the estimated proportion, we define x_i as the indicator that the outcome is observed (as we did with SRS):

$$\hat{p} = \frac{1}{n} \sum_{i=1}^{n} x_i$$

and the estimated standard error of the proportion is:

$$se(\hat{p}) = \sqrt{\frac{\hat{p}(1-\hat{p})}{n}}$$

The syntax

If we had a sample that was drawn through SRSWR and had no nonresponse or other weighting adjustments, in R, we specify this design as:

```
srswr1_des <- dat %>%
  as_survey_design()
```

where `dat` is a tibble or data.frame containing our survey data. This syntax is the same as an SRS design, except an FPC is not included. This is because when calculating a sample with replacement, the population pool to select from is no longer finite, so a correction is not needed. Therefore, with large populations where the FPC is negligible, the underlying formulas for SRS and SRSWR designs are the same.

If some post-survey adjustments were implemented and the weights are not all equal, we specify the design as:

```
srswr2_des <- dat %>%
  as_survey_design(weights = wtvar)
```

where `wtvar` is the variable for the weight of the data.

Example

The {survey} package does not include an example of SRSWR. To illustrate this design, we need to create an example. We use the APIP population data provided by the {survey} package (`apipop`) and select a sample of 200 cases using the `slice_sample()` function from the tidyverse. One of the arguments in the `slice_sample()` function is `replace`. If `replace=TRUE`, then we are conducting an SRSWR. We then calculate selection weights as the inverse of the probability of selection and call this new dataset `apisrswr`.

```
set.seed(409963)

apisrswr <- apipop %>%
  as_tibble() %>%
  slice_sample(n = 200, replace = TRUE) %>%
  select(cds, dnum, snum, dname, sname) %>%
  mutate(weight = nrow(apipop) / 200)

head(apisrswr)
```

```
## # A tibble: 6 x 6
##   cds              dnum  snum dname                     sname      weight
##   <chr>           <int> <dbl> <chr>                     <chr>       <dbl>
## 1 43696416060065    533  5348 Palo Alto Unified         Jordan (~    31.0
## 2 07618046005060    650   509 San Ramon Valley Unified  Alamo El~    31.0
## 3 19648086085674    457  2134 Montebello Unified        La Merce~    31.0
## 4 07617056003719    346   377 Knightsen Elementary      Knightse~    31.0
## 5 19650606023022    744  2351 Torrance Unified          Carr (Ev~    31.0
## 6 01611196090120      6    13 Alameda City Unified      Paden (W~    31.0
```

Because this is an SRS design with replacement, there may be duplicates in the data. It is important to keep the duplicates in the data for proper estimation. For reference, we can view the duplicates in the example data we just created.

```
apisrswr %>%
  group_by(cds) %>%
  filter(n() > 1) %>%
  arrange(cds)
```

```
## # A tibble: 4 x 6
## # Groups:   cds [2]
##   cds              dnum  snum dname                sname       weight
##   <chr>           <int> <dbl> <chr>                <chr>        <dbl>
## 1 15633216008841     41   869 Bakersfield City Elem Chipman Jun~  31.0
## 2 15633216008841     41   869 Bakersfield City Elem Chipman Jun~  31.0
## 3 39686766042782    716  4880 Stockton City Unified Tyler Skill~  31.0
## 4 39686766042782    716  4880 Stockton City Unified Tyler Skill~  31.0
```

We created a weight variable in this example data, which is the inverse of the probability of selection. We specify the sampling design for `apisrswr` as:

```
apisrswr_des <- apisrswr %>%
  as_survey_design(weights = weight)

apisrswr_des
```

```
## Independent Sampling design (with replacement)
## Called via srvyr
## Sampling variables:
##   - ids: `1`
##   - weights: weight
## Data variables:
##   - cds (chr), dnum (int), snum (dbl), dname (chr), sname (chr),
##     weight (dbl)
```

```
summary(apisrswr_des)
```

```
## Independent Sampling design (with replacement)
## Called via srvyr
## Probabilities:
##    Min. 1st Qu.  Median    Mean 3rd Qu.    Max.
##  0.0323  0.0323  0.0323  0.0323  0.0323  0.0323
## Data variables:
## [1] "cds"     "dnum"    "snum"    "dname"   "sname"   "weight"
```

In the output above, the design object and the object summary are shown. Both note that the sampling is done "with replacement" because no FPC was specified. The probabilities, which are derived from the weights, are summarized in the summary function output.

10.2.3 Stratified sampling

Stratified sampling occurs when a population is divided into mutually exclusive subpopulations (strata), and then samples are selected independently within each stratum.

- Requirements: The sampling frame must include the information to divide the population into strata for every unit.
- Advantages:
  - This design ensures sample representation in all subpopulations.
  - If the strata are correlated with survey outcomes, a stratified sample has smaller standard errors compared to a SRS sample of the same size.
  - This results in a more efficient design.
- Disadvantages: Auxiliary data may not exist to divide the sampling frame into strata, or the data may be outdated.
- Examples:
  - Example 1: A population of North Carolina residents could be stratified into urban and rural areas, and then an SRS of residents from both rural and urban areas is selected independently. This ensures there are residents from both areas in the sample.
  - Example 2: Law enforcement agencies could be stratified into the three primary general-purpose categories in the U.S.: local police, sheriff's departments, and state police. An SRS of agencies from each of the three types is then selected independently to ensure all three types of agencies are represented.

The math

Let $\bar{y}_h$ be the sample mean for stratum h, N_h be the population size of stratum h, n_h be the sample size of stratum h, and H be the total number of strata. Then, the estimate for the population mean under stratified SRS sampling is:

$$\bar{y} = \frac{1}{N} \sum_{h=1}^{H} N_h \bar{y}_h$$

and the estimate of the standard error of $\bar{y}$ is:

$$se(\bar{y}) = \sqrt{\frac{1}{N^2} \sum_{h=1}^{H} N_h^2 \frac{s_h^2}{n_h} \left(1 - \frac{n_h}{N_h}\right)}$$

where

$$s_h^2 = \frac{1}{n_h - 1} \sum_{i=1}^{n_h} (y_{i,h} - \bar{y}_h)^2$$

For estimates of proportions, let $\hat{p}_h$ be the estimated proportion in stratum h. Then, the population proportion estimate is:

$$\hat{p} = \frac{1}{N} \sum_{h=1}^{H} N_h \hat{p}_h$$

The standard error of the proportion is:

$$se(\hat{p}) = \frac{1}{N} \sqrt{\sum_{h=1}^{H} N_h^2 \frac{\hat{p}_h(1 - \hat{p}_h)}{n_h - 1} \left(1 - \frac{n_h}{N_h}\right)}$$

The syntax

In addition to the `fpc` and `weights` arguments discussed in the types above, stratified designs require the addition of the `strata` argument. For example, to specify a stratified SRS design in {srvyr} when using the FPC, that is, where the population sizes of the strata are not too large and are known, we specify the design as:

```
stsrs1_des <- dat %>%
 as_survey_design(fpc = fpcvar,
                  strata = stratavar)
```

where `fpcvar` is a variable on our data that indicates N_h for each row, and `stratavar` is a variable indicating the stratum for each row. We can omit the FPC if it is not applicable. Additionally, we can indicate the weight variable if it is present where `wtvar` is a variable on our data with a numeric weight.

```
stsrs2_des <- dat %>%
  as_survey_design(weights = wtvar,
                   strata = stratavar)
```

Example

In the example APIP data, `apistrat` is a stratified random sample, stratified by school type (`stype`) with three levels: `E` for elementary school, `M` for middle school, and `H` for high school. As with the SRS example above, we sort and select specific variables for use in printing. The data are illustrated below, including a count of the number of cases per stratum:

```
apistrat_slim <-
  apistrat %>%
  as_tibble() %>%
  arrange(dnum, snum) %>%
  select(cds, dnum, snum, dname, sname, stype, fpc, pw)

apistrat_slim %>%
  count(stype, fpc)
```

```
## # A tibble: 3 x 3
##   stype   fpc     n
##   <fct> <dbl> <int>
## 1 E      4421   100
## 2 H       755    50
## 3 M      1018    50
```

The FPC is the same for each case within each stratum. This output also shows that 100 elementary schools, 50 middle schools, and 50 high schools were sampled. It is often common for the number of units sampled from each strata to be different based on the goals of the project, or to mirror the size of each strata in the population. We specify the design as:

```
apistrat_des <- apistrat_slim %>%
  as_survey_design(
    strata = stype,
    weights = pw,
    fpc = fpc
```

)

```
apistrat_des
```

```
## Stratified Independent Sampling design
## Called via srvyr
## Sampling variables:
##    - ids: `1`
##    - strata: stype
##    - fpc: fpc
##    - weights: pw
## Data variables:
##    - cds (chr), dnum (int), snum (dbl), dname (chr), sname (chr),
##      stype (fct), fpc (dbl), pw (dbl)
```

```
summary(apistrat_des)
```

```
## Stratified Independent Sampling design
## Called via srvyr
## Probabilities:
##     Min. 1st Qu.  Median    Mean 3rd Qu.    Max.
##   0.0226  0.0226  0.0359  0.0401  0.0534  0.0662
## Stratum Sizes:
##              E  H  M
## obs        100 50 50
## design.PSU 100 50 50
## actual.PSU 100 50 50
## Population stratum sizes (PSUs):
##     E    H    M
## 4421  755 1018
## Data variables:
## [1] "cds"    "dnum"   "snum"   "dname"  "sname"  "stype"  "fpc"    "pw"
```

When printing the object, it is specified as a "Stratified Independent Sampling design," also known as a stratified SRS, and the strata variable is included. Printing the summary, we see a distribution of probabilities, as we saw with SRS; but we also see the sample and population sizes by stratum.

10.2.4 Clustered sampling

Clustered sampling occurs when a population is divided into mutually exclusive subgroups called clusters or primary sampling units (PSUs). A random selection of PSUs is sampled, and then another level of sampling is done within these clusters. There can be multiple levels of this selection. Clustered sampling is often used when a list of the entire population is not available or data collection involves interviewers needing direct contact with respondents.

- Requirements: There must be a way to divide the population into clusters. Clusters are commonly structural, such as institutions (e.g., schools, prisons) or geography (e.g., states, counties).
- Advantages:
  - Clustered sampling is advantageous when data collection is done in person, so interviewers are sent to specific sampled areas rather than completely at random across a country.
  - With clustered sampling, a list of the entire population is not necessary. For example, if sampling students, we do not need a list of all students, but only a list of all schools. Once the schools are sampled, lists of students can be obtained within the sampled schools.
- Disadvantages: Compared to a simple random sample for the same sample size, clustered samples generally have larger standard errors of estimates.
- Examples:
  - Example 1: Consider a study needing a sample of 6th-grade students in the United States. No list likely exists of all these students. However, it is more likely to obtain a list of schools that enroll 6th graders, so a study design could select a random sample of schools that enroll 6th graders. The selected schools can then provide a list of students to do a second stage of sampling where 6th-grade students are randomly sampled within each of the sampled schools. This is a one-stage sample design (the one representing the number of clusters) and is the type of design we discuss in the formulas below.
  - Example 2: Consider a study sending interviewers to households for a survey. This is a more complicated example that requires two levels of clustering (two-stage sample design) to efficiently use interviewers in geographic clusters. First, in the U.S., counties could be selected as the PSU and then census block groups within counties could be selected as the secondary sampling unit (SSU). Households could then be randomly sampled within the block groups. This type of design is popular for in-person surveys, as it reduces the travel necessary for interviewers.

The math

Consider a survey where a clusters are sampled from a population of A clusters via SRS. Within each sampled cluster, i, there are B_i units in the population, and b_i units are sampled via SRS. Let $\bar{y}_i$ be the sample mean of cluster i. Then, a ratio estimator of the population mean is:

$$\bar{y} = \frac{\sum_{i=1}^{a} B_i \bar{y}_i}{\sum_{i=1}^{a} B_i}$$

Note this is a consistent but biased estimator. Often the population size is not known, so this is a method to estimate a mean without knowing the population size. The estimated standard error of the mean is:

$$se(\bar{y}) = \frac{1}{\hat{N}} \sqrt{\left(1 - \frac{a}{A}\right)\frac{s_a^2}{a} + \frac{A}{a}\sum_{i=1}^{a}\left(1 - \frac{b_i}{B_i}\right)\frac{s_i^2}{b_i}}$$

where $\hat{N}$ is the estimated population size, s_a^2 is the between-cluster variance, and s_i^2 is the within-cluster variance.

The formula for the between-cluster variance (s_a^2) is:

$$s_a^2 = \frac{1}{a-1}\sum_{i=1}^{a}\left(\hat{y}_i - \frac{\sum_{i=1}^{a}\hat{y}_i}{a}\right)^2$$

where $\hat{y}_i = B_i\bar{y}_i$.

The formula for the within-cluster variance (s_i^2) is:

$$s_i^2 = \frac{1}{a(b_i-1)}\sum_{j=1}^{b_i}(y_{ij} - \bar{y}_i)^2$$

where y_{ij} is the outcome for sampled unit j within cluster i.

The syntax

Clustered sampling designs require the addition of the ids argument, which specifies the cluster level variable(s). To specify a two-stage clustered design without replacement, we specify the design as:

```
clus2_des <- dat %>%
  as_survey_design(weights = wtvar,
                   ids = c(PSU, SSU),
                   fpc = c(A, B))
```

where PSU and SSU are the variables indicating the PSU and SSU identifiers, and A and B are the variables indicating the population sizes for each level (i.e., A is the number of clusters, and B is the number of units within each cluster). Note that A is the same for all records, and B is the same for all records within the same cluster.

If clusters were sampled with replacement or from a very large population, the FPC is unnecessary. Additionally, only the first stage of selection is necessary regardless of whether the units were selected with replacement at any stage. The subsequent stages of selection are ignored in computation as their contribution to the variance is overpowered by the first stage (see Särndal et al. (2003)

or Wolter (2007) for a more in-depth discussion). Therefore, the two design objects specified below yield the same estimates in the end:

```
clus2ex1_des <- dat %>%
  as_survey_design(weights = wtvar,
                   ids = c(PSU, SSU))

clus2ex2_des <- dat %>%
  as_survey_design(weights = wtvar,
                   ids = PSU)
```

Note that there is one additional argument that is sometimes necessary, which is nest = TRUE. This option relabels cluster IDs to enforce nesting within strata. Sometimes, as an example, there may be a cluster 1 within each stratum, but cluster 1 in stratum 1 is a different cluster than cluster 1 in stratum 2. These are actually different clusters. This option indicates that repeated numbering does not mean it is the same cluster. If this option is not used and there are repeated cluster IDs across different strata, an error is generated.

Example

The survey package includes a two-stage cluster sample data, apiclus2, in which school districts were sampled, and then a random sample of five schools was selected within each district. For districts with fewer than five schools, all schools were sampled. School districts are identified by dnum, and schools are identified by snum. The variable fpc1 indicates how many districts there are in California (the total number of PSUs or A), and fpc2 indicates how many schools were in a given district with at least 100 students (the total number of SSUs or B). The data include a row for each school. In the data printed below, there are 757 school districts, as indicated by fpc1, and there are nine schools in District 731, one school in District 742, two schools in District 768, and so on as indicated by fpc2. For illustration purposes, the object apiclus2_slim has been created from apiclus2, which subsets the data to only the necessary columns and sorts the data.

```
apiclus2_slim <-
  apiclus2 %>%
  as_tibble() %>%
  arrange(desc(dnum), snum) %>%
  select(cds, dnum, snum, fpc1, fpc2, pw)

apiclus2_slim
```

```
## # A tibble: 126 x 6
##    cds               dnum  snum   fpc1      fpc2    pw
##    <chr>            <int> <dbl> <dbl> <int[1d]> <dbl>
##  1 47704826050942     795  5552   757         1  18.9
##  2 07618126005169     781   530   757         6  22.7
##  3 07618126005177     781   531   757         6  22.7
##  4 07618126005185     781   532   757         6  22.7
##  5 07618126005193     781   533   757         6  22.7
##  6 07618126005243     781   535   757         6  22.7
##  7 19650786023337     768  2371   757         2  18.9
##  8 19650786023345     768  2372   757         2  18.9
##  9 54722076054423     742  5898   757         1  18.9
## 10 50712906053086     731  5781   757         9  34.1
## # i 116 more rows
```

To specify this design in R, we use the following:

```
apiclus2_des <- apiclus2_slim %>%
  as_survey_design(
    ids = c(dnum, snum),
    fpc = c(fpc1, fpc2),
    weights = pw
  )

apiclus2_des
```

```
## 2 - level Cluster Sampling design
## With (40, 126) clusters.
## Called via srvyr
## Sampling variables:
##    - ids: `dnum + snum`
##    - fpc: `fpc1 + fpc2`
##    - weights: pw
## Data variables:
##    - cds (chr), dnum (int), snum (dbl), fpc1 (dbl), fpc2 (int[1d]), pw
##      (dbl)
```

```
summary(apiclus2_des)
```

```
## 2 - level Cluster Sampling design
## With (40, 126) clusters.
## Called via srvyr
## Probabilities:
##    Min. 1st Qu.  Median    Mean 3rd Qu.    Max.
## 0.00367 0.03774 0.05284 0.04239 0.05284 0.05284
## Population size (PSUs): 757
```

```
## Data variables:
## [1] "cds"  "dnum" "snum" "fpc1" "fpc2" "pw"
```

The design objects are described as "2 - level Cluster Sampling design," and include the ids (cluster), FPC, and weight variables. The summary notes that the sample includes 40 first-level clusters (PSUs), which are school districts, and 126 second-level clusters (SSUs), which are schools. Additionally, the summary includes a numeric summary of the probabilities of selection and the population size (number of PSUs) as 757.

10.3   Combining sampling methods

SRS, stratified, and clustered designs are the backbone of sampling designs, and the features are often combined in one design. Additionally, rather than using SRS for selection, other sampling mechanisms are commonly used, such as probability proportional to size (PPS), systematic sampling, or selection with unequal probabilities, which are briefly described here. In PPS sampling, a size measure is constructed for each unit (e.g., the population of the PSU or the number of occupied housing units), and units with larger size measures are more likely to be sampled. Systematic sampling is commonly used to ensure representation across a population. Units are sorted by a feature, and then every k units is selected from a random start point so the sample is spread across the population. In addition to PPS, other unequal probabilities of selection may be used. For example, in a study of establishments (e.g., businesses or public institutions) that conducts a survey every year, an establishment that recently participated (e.g., participated last year) may have a reduced chance of selection in a subsequent round to reduce the burden on the establishment. To learn more about sampling designs, refer to Valliant et al. (2013), Cox et al. (2011), Cochran (1977), and Deming (1991).

A common method of sampling is to stratify PSUs, select PSUs within the stratum using PPS selection, and then select units within the PSUs either with SRS or PPS. Reading survey documentation is an important first step in survey analysis to understand the design of the survey we are using and variables necessary to specify the design. Good documentation highlights the variables necessary to specify the design. This is often found in the user guide, methodology report, analysis guide, or technical documentation (see Chapter 3 for more details).

Example

For example, the 2017-2019 National Survey of Family Growth[1] had a stratified multi-stage area probability sample:

1. In the first stage, PSUs are counties or collections of counties and are stratified by Census region/division, size (population), and MSA status. Within each stratum, PSUs were selected via PPS.
2. In the second stage, neighborhoods were selected within the sampled PSUs using PPS selection.
3. In the third stage, housing units were selected within the sampled neighborhoods.
4. In the fourth stage, a person was randomly chosen among eligible persons within the selected housing units using unequal probabilities based on the person's age and sex.

The public use file does not include all these levels of selection and instead has pseudo-strata and pseudo-clusters, which are the variables used in R to specify the design. As specified on page 4 of the documentation, the stratum variable is SEST, the cluster variable is SECU, and the weight variable is WGT2017_2019. Thus, to specify this design in R, we use the following syntax:

```
nsfg_des <- nsfgdata %>%
  as_survey_design(ids = SECU,
                   strata = SEST,
                   weights = WGT2017_2019)
```

10.4 Replicate weights

Replicate weights are often included on analysis files instead of, or in addition to, the design variables (strata and PSUs). Replicate weights are used as another method to estimate variability. Often, researchers choose to use replicate weights to avoid publishing design variables (strata or clustering variables) as a measure to reduce the risk of disclosure. There are several types of replicate weights, including balanced repeated replication (BRR), Fay's BRR, jackknife, and bootstrap methods. An overview of the process for using replicate weights is as follows:

---

[1]https://www.cdc.gov/nchs/data/nsfg/NSFG-2017-2019-Sample-Design-Documentation-508.pdf

1. Divide the sample into subsample replicates that mirror the design of the sample
2. Calculate weights for each replicate using the same procedures for the full-sample weight (i.e., nonresponse and post-stratification)
3. Calculate estimates for each replicate using the same method as the full-sample estimate
4. Calculate the estimated variance, which is proportional to the variance of the replicate estimates

The different types of replicate weights largely differ between step 1 (how the sample is divided into subsamples) and step 4 (which multiplication factors, scales, are used to multiply the variance). The general format for the standard error is:

$$\sqrt{\alpha \sum_{r=1}^{R} \alpha_r (\hat{\theta}_r - \hat{\theta})^2}$$

where R is the number of replicates, α is a constant that depends on the replication method, α_r is a factor associated with each replicate, $\hat{\theta}$ is the weighted estimate based on the full sample, and $\hat{\theta}_r$ is the weighted estimate of θ based on the rth replicate.

To create the design object for surveys with replicate weights, we use as_survey_rep() instead of as_survey_design(), which we use for the common sampling designs in the sections above.

10.4.1   Balanced Repeated Replication method

The balanced repeated replication (BRR) method requires a stratified sample design with two PSUs in each stratum. Each replicate is constructed by deleting one PSU per stratum using a Hadamard matrix. For the PSU that is included, the weight is generally multiplied by two but may have other adjustments, such as post-stratification. A Hadamard matrix is a special square matrix with entries of +1 or −1 with mutually orthogonal rows. Hadamard matrices must have one row, two rows, or a multiple of four rows. The size of the Hadamard matrix is determined by the first multiple of 4 greater than or equal to the number of strata. For example, if a survey had seven strata, the Hadamard matrix would be an 8 × 8 matrix. Additionally, a survey with eight strata would also have an 8 × 8 Hadamard matrix. The columns in the matrix specify the strata, and the rows specify the replicate. In each replicate (row), a +1 means to use the first PSU, and a −1 means to use the second PSU in the estimate. For example, here is a 4 × 4 Hadamard matrix:

$$
\begin{array}{cccc}
+1 & +1 & +1 & +1 \\
+1 & -1 & +1 & -1 \\
+1 & +1 & -1 & -1 \\
+1 & -1 & -1 & +1
\end{array}
$$

In the first replicate (row), all the values are +1; so in each stratum, the first PSU would be used in the estimate. In the second replicate, the first PSU would be used in strata 1 and 3, while the second PSU would be used in strata 2 and 4. In the third replicate, the first PSU would be used in strata 1 and 2, while the second PSU would be used in strata 3 and 4. Finally, in the fourth replicate, the first PSU would be used in strata 1 and 4, while the second PSU would be used in strata 2 and 3. For more information about Hadamard matrices, see Wolter (2007). Note that supplied BRR weights from a data provider already incorporate this adjustment, and the {survey} package generates the Hadamard matrix, if necessary, for calculating BRR weights; so an analyst does not need to create or provide the matrix.

The math

A weighted estimate for the full sample is calculated as $\hat{\theta}$, and then a weighted estimate for each replicate is calculated as $\hat{\theta}_r$ for R replicates. Using the generic notation above, $\alpha = \frac{1}{R}$ and $\alpha_r = 1$ for each r. The standard error of the estimate is calculated as follows:

$$
se(\hat{\theta}) = \sqrt{\frac{1}{R} \sum_{r=1}^{R} \left(\hat{\theta}_r - \hat{\theta} \right)^2}
$$

Specifying replicate weights in R requires specifying the type of replicate weights, the main weight variable, the replicate weight variables, and other options. One of the key options is for the mean squared error (MSE). If mse=TRUE, variances are computed around the point estimate ($\hat{\theta}$); whereas if mse=FALSE, variances are computed around the mean of the replicates ($\bar{\theta}$) instead, which looks like this:

$$
se(\hat{\theta}) = \sqrt{\frac{1}{R} \sum_{r=1}^{R} \left(\hat{\theta}_r - \bar{\theta} \right)^2}
$$

where

$$
\bar{\theta} = \frac{1}{R} \sum_{r=1}^{R} \hat{\theta}_r
$$

The default option for mse is to use the global option of "survey.replicates.mse," which is set to FALSE initially unless a user changes it. To determine if mse should be set to TRUE or FALSE, read the survey documentation. If there is no indication in the survey documentation for BRR, we recommend setting mse to TRUE, as this is the default in other software (e.g., SAS, SUDAAN).

The syntax

Replicate weights generally come in groups and are sequentially numbered, such as PWGTP1, PWGTP2, ..., PWGTP80 for the person weights in the American Community Survey (ACS) (U.S. Census Bureau, 2021) or BRRWT1, BRRWT2, ..., BRRWT96 in the 2015 Residential Energy Consumption Survey (RECS) (U.S. Energy Information Administration, 2017). This makes it easy to use some of the tidy selection[2] functions in R.

To specify a BRR design, we need to specify the weight variable (weights), the replicate weight variables (repweights), the type of replicate weights as BRR (type = BRR), and whether the mean squared error should be used (mse = TRUE) or not (mse = FALSE). For example, if a dataset had WT0 for the main weight and had 20 BRR weights indicated WT1, WT2, ..., WT20, we can use the following syntax (both are equivalent):

```
brr_des <- dat %>%
  as_survey_rep(weights = WT0,
                repweights = all_of(str_c("WT", 1:20)),
                type = "BRR",
                mse = TRUE)
```

```
brr_des <- dat %>%
  as_survey_rep(weights = WT0,
                repweights = num_range("WT", 1:20),
                type = "BRR",
                mse = TRUE)
```

If a dataset had WT for the main weight and had 20 BRR weights indicated REPWT1, REPWT2, ..., REPWT20, we can use the following syntax (both are equivalent):

```
brr_des <- dat %>%
  as_survey_rep(weights = WT,
                repweights = all_of(str_c("REPWT", 1:20)),
                type = "BRR",
                mse = TRUE)
```

```
brr_des <- dat %>%
  as_survey_rep(weights = WT,
```

[2]https://dplyr.tidyverse.org/reference/dplyr_tidy_select.html

```
                    repweights = starts_with("REPWT"),
                    type = "BRR",
                    mse = TRUE)
```

If the replicate weight variables are in the file consecutively, we can also use the following syntax:

```
brr_des <- dat %>%
  as_survey_rep(weights = WT,
                repweights = REPWT1:REPWT20,
                type = "BRR",
                mse = TRUE)
```

Typically, each replicate weight sums to a value similar to the main weight, as both the replicate weights and the main weight are supposed to provide population estimates. Rarely, an alternative method is used where the replicate weights have values of 0 or 2 in the case of BRR weights. This would be indicated in the documentation (see Chapter 3 for more information on reading documentation). In this case, the replicate weights are not combined, and the option combined_weights = FALSE should be indicated, as the default value for this argument is TRUE. This specific syntax is shown below:

```
brr_des <- dat %>%
  as_survey_rep(weights = WT,
                repweights = starts_with("REPWT"),
                type = "BRR",
                combined_weights = FALSE,
                mse = TRUE)
```

Example

The {survey} package includes a data example from section 12.2 of Levy and Lemeshow (2013). In this fictional data, two out of five ambulance stations were sampled from each of three emergency service areas (ESAs); thus BRR weights are appropriate with two PSUs (stations) sampled in each stratum (ESA). In the code below, we create BRR weights as was done by Levy and Lemeshow (2013).

```
scdbrr <- scd %>%
  as_tibble() %>%
  mutate(
    wt = 5 / 2,
    rep1 = 2 * c(1, 0, 1, 0, 1, 0),
```

```
    rep2 = 2 * c(1, 0, 0, 1, 0, 1),
    rep3 = 2 * c(0, 1, 1, 0, 0, 1),
    rep4 = 2 * c(0, 1, 0, 1, 1, 0)
  )

scdbrr
```

```
## # A tibble: 6 x 9
##      ESA ambulance arrests alive    wt  rep1  rep2  rep3  rep4
##    <int>     <int>   <dbl> <dbl> <dbl> <dbl> <dbl> <dbl> <dbl>
## 1      1         1     120    25   2.5     2     2     0     0
## 2      1         2      78    24   2.5     0     0     2     2
## 3      2         1     185    30   2.5     2     0     2     0
## 4      2         2     228    49   2.5     0     2     0     2
## 5      3         1     670    80   2.5     2     0     0     2
## 6      3         2     530    70   2.5     0     2     2     0
```

To specify the BRR weights, we use the following syntax:

```
scdbrr_des <- scdbrr %>%
  as_survey_rep(
    type = "BRR",
    repweights = starts_with("rep"),
    combined_weights = FALSE,
    weight = wt
  )

scdbrr_des
```

```
## Call: Called via srvyr
## Balanced Repeated Replicates with 4 replicates.
## Sampling variables:
##   - repweights: `rep1 + rep2 + rep3 + rep4`
##   - weights: wt
## Data variables:
##   - ESA (int), ambulance (int), arrests (dbl), alive (dbl), wt (dbl),
##     rep1 (dbl), rep2 (dbl), rep3 (dbl), rep4 (dbl)
```

```
summary(scdbrr_des)
```

```
## Call: Called via srvyr
## Balanced Repeated Replicates with 4 replicates.
## Sampling variables:
##   - repweights: `rep1 + rep2 + rep3 + rep4`
##   - weights: wt
```

```
## Data variables:
##   - ESA (int), ambulance (int), arrests (dbl), alive (dbl), wt (dbl),
##     rep1 (dbl), rep2 (dbl), rep3 (dbl), rep4 (dbl)
## Variables:
## [1] "ESA"      "ambulance" "arrests"  "alive"    "wt"
## [6] "rep1"     "rep2"      "rep3"     "rep4"
```

Note that `combined_weights` was specified as `FALSE` because these weights are simply specified as 0 and 2 and do not incorporate the overall weight. When printing the object, the type of replication is noted as Balanced Repeated Replicates, and the replicate weights and the weight variable are specified. Additionally, the summary lists the variables included in the data and design object.

10.4.2 Fay's BRR method

Fay's BRR method for replicate weights is similar to the BRR method in that it uses a Hadamard matrix to construct replicate weights. However, rather than deleting PSUs for each replicate, with Fay's BRR, half of the PSUs have a replicate weight, which is the main weight multiplied by ρ, and the other half have the main weight multiplied by $(2 - \rho)$, where $0 \leq \rho < 1$. Note that when $\rho = 0$, this is equivalent to the standard BRR weights, and as ρ becomes closer to 1, this method is more similar to jackknife discussed in Section 10.4.3. To obtain the value of ρ, it is necessary to read the survey documentation (see Chapter 3).

The math

The standard error estimate for $\hat{\theta}$ is slightly different than the BRR, due to the addition of the multiplier of ρ. Using the generic notation above, $\alpha = \frac{1}{R(1-\rho)^2}$ and $\alpha_r = 1$ for all r. The standard error is calculated as:

$$se(\hat{\theta}) = \sqrt{\frac{1}{R(1 - \rho)^2} \sum_{r=1}^{R} \left(\hat{\theta}_r - \hat{\theta}\right)^2}$$

The syntax

The syntax is very similar for BRR and Fay's BRR. To specify a Fay's BRR design, we need to specify the weight variable (`weights`), the replicate weight variables (`repweights`), the type of replicate weights as Fay's BRR (`type = Fay`), whether the mean squared error should be used (`mse = TRUE`) or not (`mse = FALSE`), and Fay's multiplier (`rho`). For example, if a dataset had WT0 for the main weight and had 20 BRR weights indicated as WT1, WT2, ..., WT20, and Fay's multiplier is 0.3, we use the following syntax:

```
fay_des <- dat %>%
  as_survey_rep(weights = WT0,
                repweights = num_range("WT", 1:20),
                type = "Fay",
                mse = TRUE,
                rho = 0.3)
```

Example

The 2015 RECS (U.S. Energy Information Administration, 2017) uses Fay's
BRR weights with the final weight as NWEIGHT and replicate weights as
BRRWT1 - BRRWT96, and the documentation specifies a Fay's multiplier of
0.5. On the file, DOEID is a unique identifier for each respondent, TOTALDOL
is the total energy cost, TOTSQFT_EN is the total square footage of the
residence, and REGOINC is the census region. We use the 2015 RECS data
from the {srvyrexploR} package that provides data for this book (see the
Prerequisites box at the beginning of this chapter). To specify the design for
the recs_2015 data, we use the following syntax:

```
recs_2015_des <- recs_2015 %>%
  as_survey_rep(
    weights = NWEIGHT,
    repweights = BRRWT1:BRRWT96,
    type = "Fay",
    rho = 0.5,
    mse = TRUE,
    variables = c(DOEID, TOTALDOL, TOTSQFT_EN, REGIONC)
  )
```

```
recs_2015_des
```

```
## Call: Called via srvyr
## Fay's variance method (rho= 0.5 ) with 96 replicates and MSE variances.
## Sampling variables:
##   - repweights: `BRRWT1 + BRRWT2 + BRRWT3 + BRRWT4 + BRRWT5 + BRRWT6
##       + BRRWT7 + BRRWT8 + BRRWT9 + BRRWT10 + BRRWT11 + BRRWT12 +
##       BRRWT13 + BRRWT14 + BRRWT15 + BRRWT16 + BRRWT17 + BRRWT18 +
##       BRRWT19 + BRRWT20 + BRRWT21 + BRRWT22 + BRRWT23 + BRRWT24 +
##       BRRWT25 + BRRWT26 + BRRWT27 + BRRWT28 + BRRWT29 + BRRWT30 +
##       BRRWT31 + BRRWT32 + BRRWT33 + BRRWT34 + BRRWT35 + BRRWT36 +
##       BRRWT37 + BRRWT38 + BRRWT39 + BRRWT40 + BRRWT41 + BRRWT42 +
##       BRRWT43 + BRRWT44 + BRRWT45 + BRRWT46 + BRRWT47 + BRRWT48 +
##       BRRWT49 + BRRWT50 + BRRWT51 + BRRWT52 + BRRWT53 + BRRWT54 +
##       BRRWT55 + BRRWT56 + BRRWT57 + BRRWT58 + BRRWT59 + BRRWT60 +
##       BRRWT61 + BRRWT62 + BRRWT63 + BRRWT64 + BRRWT65 + BRRWT66 +
```

```
##        BRRWT67 + BRRWT68 + BRRWT69 + BRRWT70 + BRRWT71 + BRRWT72 +
##        BRRWT73 + BRRWT74 + BRRWT75 + BRRWT76 + BRRWT77 + BRRWT78 +
##        BRRWT79 + BRRWT80 + BRRWT81 + BRRWT82 + BRRWT83 + BRRWT84 +
##        BRRWT85 + BRRWT86 + BRRWT87 + BRRWT88 + BRRWT89 + BRRWT90 +
##        BRRWT91 + BRRWT92 + BRRWT93 + BRRWT94 + BRRWT95 + BRRWT96`
##    - weights: NWEIGHT
## Data variables:
##    - DOEID (dbl), TOTALDOL (dbl), TOTSQFT_EN (dbl), REGIONC (dbl)
```

```
summary(recs_2015_des)
```

```
## Call: Called via srvyr
## Fay's variance method (rho= 0.5 ) with 96 replicates and MSE variances.
## Sampling variables:
##    - repweights: `BRRWT1 + BRRWT2 + BRRWT3 + BRRWT4 + BRRWT5 + BRRWT6
##      + BRRWT7 + BRRWT8 + BRRWT9 + BRRWT10 + BRRWT11 + BRRWT12 +
##        BRRWT13 + BRRWT14 + BRRWT15 + BRRWT16 + BRRWT17 + BRRWT18 +
##        BRRWT19 + BRRWT20 + BRRWT21 + BRRWT22 + BRRWT23 + BRRWT24 +
##        BRRWT25 + BRRWT26 + BRRWT27 + BRRWT28 + BRRWT29 + BRRWT30 +
##        BRRWT31 + BRRWT32 + BRRWT33 + BRRWT34 + BRRWT35 + BRRWT36 +
##        BRRWT37 + BRRWT38 + BRRWT39 + BRRWT40 + BRRWT41 + BRRWT42 +
##        BRRWT43 + BRRWT44 + BRRWT45 + BRRWT46 + BRRWT47 + BRRWT48 +
##        BRRWT49 + BRRWT50 + BRRWT51 + BRRWT52 + BRRWT53 + BRRWT54 +
##        BRRWT55 + BRRWT56 + BRRWT57 + BRRWT58 + BRRWT59 + BRRWT60 +
##        BRRWT61 + BRRWT62 + BRRWT63 + BRRWT64 + BRRWT65 + BRRWT66 +
##        BRRWT67 + BRRWT68 + BRRWT69 + BRRWT70 + BRRWT71 + BRRWT72 +
##        BRRWT73 + BRRWT74 + BRRWT75 + BRRWT76 + BRRWT77 + BRRWT78 +
##        BRRWT79 + BRRWT80 + BRRWT81 + BRRWT82 + BRRWT83 + BRRWT84 +
##        BRRWT85 + BRRWT86 + BRRWT87 + BRRWT88 + BRRWT89 + BRRWT90 +
##        BRRWT91 + BRRWT92 + BRRWT93 + BRRWT94 + BRRWT95 + BRRWT96`
##    - weights: NWEIGHT
## Data variables:
##    - DOEID (dbl), TOTALDOL (dbl), TOTSQFT_EN (dbl), REGIONC (dbl)
## Variables:
## [1] "DOEID"      "TOTALDOL"   "TOTSQFT_EN" "REGIONC"
```

In specifying the design, the `variables` option was also used to include which variables might be used in analyses. This is optional but can make our object smaller and easier to work with. When printing the design object or looking at the summary, the replicate weight type is re-iterated as `Fay's variance method (rho= 0.5) with 96 replicates and MSE variances`, and the variables are included. No weight or probability summary is included in this output, as we have seen in some other design objects.

10.4.3 Jackknife method

There are three jackknife estimators implemented in {srvyr}: jackknife 1 (JK1), jackknife n (JKn), and jackknife 2 (JK2). The JK1 method can be used for unstratified designs, and replicates are created by removing one PSU at a time so the number of replicates is the same as the number of PSUs. If there is no clustering, then the PSU is the ultimate sampling unit (e.g., students).

The JKn method is used for stratified designs and requires two or more PSUs per stratum. In this case, each replicate is created by deleting one PSU from a single stratum, so the number of replicates is the number of total PSUs across all strata. The JK2 method is a special case of JKn when there are exactly 2 PSUs sampled per stratum. For variance estimation, we also need to specify the scaling constants.

The math

Using the generic notation above, $\alpha = \frac{R-1}{R}$ and $\alpha_r = 1$ for all r. For the JK1 method, the standard error estimate for $\hat{\theta}$ is calculated as:

$$se(\hat{\theta}) = \sqrt{\frac{R-1}{R} \sum_{r=1}^{R} \left(\hat{\theta}_r - \hat{\theta}\right)^2}$$

The JKn method is a bit more complex, but the coefficients are generally provided with restricted and public-use files. For each replicate, one stratum has a PSU removed, and the weights are adjusted by $n_h/(n_h - 1)$ where n_h is the number of PSUs in stratum h. The coefficients in other strata are set to 1. Denote the coefficient that results from this process for replicate r as α_r, then the standard error estimate for $\hat{\theta}$ is calculated as:

$$se(\hat{\theta}) = \sqrt{\sum_{r=1}^{R} \alpha_r \left(\hat{\theta}_r - \hat{\theta}\right)^2}$$

The syntax

To specify the jackknife method, we use the survey documentation to understand the type of jackknife (1, n, or 2) and the multiplier. In the syntax, we need to specify the weight variable (`weights`), the replicate weight variables (`repweights`), the type of replicate weights as jackknife 1 (`type = "JK1"`), n (`type = "JKN"`), or 2 (`type = "JK2"`), whether the mean squared error should be used (`mse = TRUE`) or not (`mse = FALSE`), and the multiplier (`scale`). For example, if the survey is a jackknife 1 method with a multiplier of $\alpha_r = (R-1)/R = 19/20 = 0.95$, the dataset has WT0 for the main weight and 20 replicate weights indicated as WT1, WT2, ..., WT20, we use the following syntax:

```
jk1_des <- dat %>%
  as_survey_rep(
```

```
    weights = WT0,
    repweights = num_range("WT", 1:20),
    type = "JK1",
    mse = TRUE,
    scale = 0.95
)
```

For a jackknife n method, we need to specify the multiplier for all replicates. In this case, we use the `rscales` argument to specify each one. The documentation provides details on what the multipliers (α_r) are, and they may be the same for all replicates. For example, consider a case where $\alpha_r = 0.1$ for all replicates, and the dataset had WT0 for the main weight and had 20 replicate weights indicated as WT1, WT2, ..., WT20. We specify the type as `type = "JKN"`, and the multiplier as `rscales=rep(0.1,20)`:

```
jkn_des <- dat %>%
  as_survey_rep(
    weights = WT0,
    repweights = num_range("WT", 1:20),
    type = "JKN",
    mse = TRUE,
    rscales = rep(0.1, 20)
  )
```

Example

The 2020 RECS (U.S. Energy Information Administration, 2023c) uses jack-knife weights with the final weight as NWEIGHT and replicate weights as NWEIGHT1 - NWEIGHT60 with a scale of $(R-1)/R = 59/60$. On the file, DOEID is a unique identifier for each respondent, TOTALDOL is the total cost of energy, TOTSQFT_EN is the total square footage of the residence, and REGOINC is the census region. We use the 2020 RECS data from the {srvyrexploR} package that provides data for this book (see the Prerequisites box at the beginning of this chapter).

To specify this design, we use the following syntax:

```
recs_des <- recs_2020 %>%
  as_survey_rep(
    weights = NWEIGHT,
    repweights = NWEIGHT1:NWEIGHT60,
    type = "JK1",
    scale = 59 / 60,
    mse = TRUE,
```

```
    variables = c(DOEID, TOTALDOL, TOTSQFT_EN, REGIONC)
 )
```

```
recs_des
```

```
## Call: Called via srvyr
## Unstratified cluster jacknife (JK1) with 60 replicates and MSE variances.
## Sampling variables:
##    - repweights: `NWEIGHT1 + NWEIGHT2 + NWEIGHT3 + NWEIGHT4 + NWEIGHT5
##      + NWEIGHT6 + NWEIGHT7 + NWEIGHT8 + NWEIGHT9 + NWEIGHT10 +
##      NWEIGHT11 + NWEIGHT12 + NWEIGHT13 + NWEIGHT14 + NWEIGHT15 +
##      NWEIGHT16 + NWEIGHT17 + NWEIGHT18 + NWEIGHT19 + NWEIGHT20 +
##      NWEIGHT21 + NWEIGHT22 + NWEIGHT23 + NWEIGHT24 + NWEIGHT25 +
##      NWEIGHT26 + NWEIGHT27 + NWEIGHT28 + NWEIGHT29 + NWEIGHT30 +
##      NWEIGHT31 + NWEIGHT32 + NWEIGHT33 + NWEIGHT34 + NWEIGHT35 +
##      NWEIGHT36 + NWEIGHT37 + NWEIGHT38 + NWEIGHT39 + NWEIGHT40 +
##      NWEIGHT41 + NWEIGHT42 + NWEIGHT43 + NWEIGHT44 + NWEIGHT45 +
##      NWEIGHT46 + NWEIGHT47 + NWEIGHT48 + NWEIGHT49 + NWEIGHT50 +
##      NWEIGHT51 + NWEIGHT52 + NWEIGHT53 + NWEIGHT54 + NWEIGHT55 +
##      NWEIGHT56 + NWEIGHT57 + NWEIGHT58 + NWEIGHT59 + NWEIGHT60`
##    - weights: NWEIGHT
## Data variables:
##    - DOEID (dbl), TOTALDOL (dbl), TOTSQFT_EN (dbl), REGIONC (chr)
```

```
summary(recs_des)
```

```
## Call: Called via srvyr
## Unstratified cluster jacknife (JK1) with 60 replicates and MSE variances.
## Sampling variables:
##    - repweights: `NWEIGHT1 + NWEIGHT2 + NWEIGHT3 + NWEIGHT4 + NWEIGHT5
##      + NWEIGHT6 + NWEIGHT7 + NWEIGHT8 + NWEIGHT9 + NWEIGHT10 +
##      NWEIGHT11 + NWEIGHT12 + NWEIGHT13 + NWEIGHT14 + NWEIGHT15 +
##      NWEIGHT16 + NWEIGHT17 + NWEIGHT18 + NWEIGHT19 + NWEIGHT20 +
##      NWEIGHT21 + NWEIGHT22 + NWEIGHT23 + NWEIGHT24 + NWEIGHT25 +
##      NWEIGHT26 + NWEIGHT27 + NWEIGHT28 + NWEIGHT29 + NWEIGHT30 +
##      NWEIGHT31 + NWEIGHT32 + NWEIGHT33 + NWEIGHT34 + NWEIGHT35 +
##      NWEIGHT36 + NWEIGHT37 + NWEIGHT38 + NWEIGHT39 + NWEIGHT40 +
##      NWEIGHT41 + NWEIGHT42 + NWEIGHT43 + NWEIGHT44 + NWEIGHT45 +
##      NWEIGHT46 + NWEIGHT47 + NWEIGHT48 + NWEIGHT49 + NWEIGHT50 +
##      NWEIGHT51 + NWEIGHT52 + NWEIGHT53 + NWEIGHT54 + NWEIGHT55 +
##      NWEIGHT56 + NWEIGHT57 + NWEIGHT58 + NWEIGHT59 + NWEIGHT60`
##    - weights: NWEIGHT
## Data variables:
##    - DOEID (dbl), TOTALDOL (dbl), TOTSQFT_EN (dbl), REGIONC (chr)
## Variables:
## [1] "DOEID"      "TOTALDOL"   "TOTSQFT_EN" "REGIONC"
```

When printing the design object or looking at the summary, the replicate weight type is reiterated as Unstratified cluster jacknife (JK1) with 60 replicates and MSE variances, and the variables are included. No weight or probability summary is included.

10.4.4 Bootstrap method

In bootstrap resampling, replicates are created by selecting random samples of the PSUs with replacement (SRSWR). If there are A PSUs in the sample, then each replicate is created by selecting a random sample of A PSUs with replacement. Each replicate is created independently, and the weights for each replicate are adjusted to reflect the population, generally using the same method as how the analysis weight was adjusted.

The math

A weighted estimate for the full sample is calculated as $\hat{\theta}$, and then a weighted estimate for each replicate is calculated as $\hat{\theta}_r$ for R replicates. Then the standard error of the estimate is calculated as follows:

$$se(\hat{\theta}) = \sqrt{\alpha \sum_{r=1}^{R} \left(\hat{\theta}_r - \hat{\theta}\right)^2}$$

where α is the scaling constant. Note that the scaling constant (α) is provided in the survey documentation, as there are many types of bootstrap methods that generate custom scaling constants.

The syntax

To specify a bootstrap method, we need to specify the weight variable (weights), the replicate weight variables (repweights), the type of replicate weights as bootstrap (type = "bootstrap"), whether the mean squared error should be used (mse = TRUE) or not (mse = FALSE), and the multiplier (scale). For example, if a dataset had WT0 for the main weight, 20 bootstrap weights indicated WT1, WT2, ..., WT20, and a multiplier of $\alpha = .02$, we use the following syntax:

```
bs_des <- dat %>%
  as_survey_rep(
    weights = WT0,
    repweights = num_range("WT", 1:20),
    type = "bootstrap",
    mse = TRUE,
```

```
    scale = .02
)
```

Example

Returning to the APIP example, we are going to create a dataset with bootstrap
weights to use as an example. In this example, we construct a one-cluster
design with 50 replicate weights[3].

```
apiclus1_slim <-
    apiclus1 %>%
    as_tibble() %>%
    arrange(dnum) %>%
    select(cds, dnum, fpc, pw)

set.seed(662152)
apibw <-
    bootweights(
        psu = apiclus1_slim$dnum,
        strata = rep(1, nrow(apiclus1_slim)),
        fpc = apiclus1_slim$fpc,
        replicates = 50
    )

bwmata <-
    apibw$repweights$weights[apibw$repweights$index, ] * apiclus1_slim$pw

apiclus1_slim <- bwmata %>%
    as.data.frame() %>%
    set_names(str_c("pw", 1:50)) %>%
    cbind(apiclus1_slim) %>%
    as_tibble() %>%
    select(cds, dnum, fpc, pw, everything())

apiclus1_slim
```

---

[3]We provide the code here to replicate this example but are not focusing on the creation
of the weights, as that is outside the scope of this book. We recommend referencing Wolter
(2007) for more information on creating bootstrap weights.

```
## # A tibble: 183 x 54
##    cds      dnum  fpc    pw    pw1   pw2   pw3   pw4   pw5   pw6   pw7
##    <chr>   <int> <dbl> <dbl> <dbl> <dbl> <dbl> <dbl> <dbl> <dbl> <dbl>
##  1 436937~   61   757  33.8  33.8     0     0  33.8     0  33.8     0
##  2 436937~   61   757  33.8  33.8     0     0  33.8     0  33.8     0
##  3 436937~   61   757  33.8  33.8     0     0  33.8     0  33.8     0
##  4 436937~   61   757  33.8  33.8     0     0  33.8     0  33.8     0
##  5 436937~   61   757  33.8  33.8     0     0  33.8     0  33.8     0
##  6 436937~   61   757  33.8  33.8     0     0  33.8     0  33.8     0
##  7 436937~   61   757  33.8  33.8     0     0  33.8     0  33.8     0
##  8 436937~   61   757  33.8  33.8     0     0  33.8     0  33.8     0
##  9 436937~   61   757  33.8  33.8     0     0  33.8     0  33.8     0
## 10 436937~   61   757  33.8  33.8     0     0  33.8     0  33.8     0
## # i 173 more rows
## # i 43 more variables: pw8 <dbl>, pw9 <dbl>, pw10 <dbl>, pw11 <dbl>,
## #   pw12 <dbl>, pw13 <dbl>, pw14 <dbl>, pw15 <dbl>, pw16 <dbl>,
## #   pw17 <dbl>, pw18 <dbl>, pw19 <dbl>, pw20 <dbl>, pw21 <dbl>,
## #   pw22 <dbl>, pw23 <dbl>, pw24 <dbl>, pw25 <dbl>, pw26 <dbl>,
## #   pw27 <dbl>, pw28 <dbl>, pw29 <dbl>, pw30 <dbl>, pw31 <dbl>,
## #   pw32 <dbl>, pw33 <dbl>, pw34 <dbl>, pw35 <dbl>, pw36 <dbl>, ...
```

The output of `apiclus1_slim` includes the same variables we have seen in other APIP examples (see Table 10.1), but now it additionally includes bootstrap weights `pw1`, ..., `pw50`. When creating the survey design object, we use the bootstrap weights as the replicate weights. Additionally, with replicate weights we need to include the scale (α). For this example, we created:

$$\alpha = \frac{A}{(A-1)(R-1)} = \frac{15}{(15-1)*(50-1)} = 0.02186589$$

where A is the average number of PSUs per stratum, and R is the number of replicates. There is only 1 stratum and the number of clusters/PSUs is 15 so $A = 15$. Using this information, we specify the design object as:

```
api1_bs_des <- apiclus1_slim %>%
  as_survey_rep(
    weights = pw,
    repweights = pw1:pw50,
    type = "bootstrap",
    scale = 0.02186589,
    mse = TRUE
  )

api1_bs_des
```

```
## Call: Called via srvyr
## Survey bootstrap with 50 replicates and MSE variances.
## Sampling variables:
##   - repweights: `pw1 + pw2 + pw3 + pw4 + pw5 + pw6 + pw7 + pw8 + pw9
##     + pw10 + pw11 + pw12 + pw13 + pw14 + pw15 + pw16 + pw17 + pw18 +
##     pw19 + pw20 + pw21 + pw22 + pw23 + pw24 + pw25 + pw26 + pw27 +
##     pw28 + pw29 + pw30 + pw31 + pw32 + pw33 + pw34 + pw35 + pw36 +
##     pw37 + pw38 + pw39 + pw40 + pw41 + pw42 + pw43 + pw44 + pw45 +
##     pw46 + pw47 + pw48 + pw49 + pw50`
##   - weights: pw
## Data variables:
##   - cds (chr), dnum (int), fpc (dbl), pw (dbl), pw1 (dbl), pw2 (dbl),
##     pw3 (dbl), pw4 (dbl), pw5 (dbl), pw6 (dbl), pw7 (dbl), pw8 (dbl),
##     pw9 (dbl), pw10 (dbl), pw11 (dbl), pw12 (dbl), pw13 (dbl), pw14
##     (dbl), pw15 (dbl), pw16 (dbl), pw17 (dbl), pw18 (dbl), pw19
##     (dbl), pw20 (dbl), pw21 (dbl), pw22 (dbl), pw23 (dbl), pw24
##     (dbl), pw25 (dbl), pw26 (dbl), pw27 (dbl), pw28 (dbl), pw29
##     (dbl), pw30 (dbl), pw31 (dbl), pw32 (dbl), pw33 (dbl), pw34
##     (dbl), pw35 (dbl), pw36 (dbl), pw37 (dbl), pw38 (dbl), pw39
##     (dbl), pw40 (dbl), pw41 (dbl), pw42 (dbl), pw43 (dbl), pw44
##     (dbl), pw45 (dbl), pw46 (dbl), pw47 (dbl), pw48 (dbl), pw49
##     (dbl), pw50 (dbl)
```

summary(api1_bs_des)

```
## Call: Called via srvyr
## Survey bootstrap with 50 replicates and MSE variances.
## Sampling variables:
##   - repweights: `pw1 + pw2 + pw3 + pw4 + pw5 + pw6 + pw7 + pw8 + pw9
##     + pw10 + pw11 + pw12 + pw13 + pw14 + pw15 + pw16 + pw17 + pw18 +
##     pw19 + pw20 + pw21 + pw22 + pw23 + pw24 + pw25 + pw26 + pw27 +
##     pw28 + pw29 + pw30 + pw31 + pw32 + pw33 + pw34 + pw35 + pw36 +
##     pw37 + pw38 + pw39 + pw40 + pw41 + pw42 + pw43 + pw44 + pw45 +
##     pw46 + pw47 + pw48 + pw49 + pw50`
##   - weights: pw
## Data variables:
##   - cds (chr), dnum (int), fpc (dbl), pw (dbl), pw1 (dbl), pw2 (dbl),
##     pw3 (dbl), pw4 (dbl), pw5 (dbl), pw6 (dbl), pw7 (dbl), pw8 (dbl),
##     pw9 (dbl), pw10 (dbl), pw11 (dbl), pw12 (dbl), pw13 (dbl), pw14
##     (dbl), pw15 (dbl), pw16 (dbl), pw17 (dbl), pw18 (dbl), pw19
##     (dbl), pw20 (dbl), pw21 (dbl), pw22 (dbl), pw23 (dbl), pw24
##     (dbl), pw25 (dbl), pw26 (dbl), pw27 (dbl), pw28 (dbl), pw29
##     (dbl), pw30 (dbl), pw31 (dbl), pw32 (dbl), pw33 (dbl), pw34
##     (dbl), pw35 (dbl), pw36 (dbl), pw37 (dbl), pw38 (dbl), pw39
##     (dbl), pw40 (dbl), pw41 (dbl), pw42 (dbl), pw43 (dbl), pw44
##     (dbl), pw45 (dbl), pw46 (dbl), pw47 (dbl), pw48 (dbl), pw49
##     (dbl), pw50 (dbl)
## Variables:
```

```
##  [1] "cds"   "dnum"  "fpc"   "pw"    "pw1"   "pw2"   "pw3"   "pw4"   "pw5"
## [10] "pw6"   "pw7"   "pw8"   "pw9"   "pw10"  "pw11"  "pw12"  "pw13"  "pw14"
## [19] "pw15"  "pw16"  "pw17"  "pw18"  "pw19"  "pw20"  "pw21"  "pw22"  "pw23"
## [28] "pw24"  "pw25"  "pw26"  "pw27"  "pw28"  "pw29"  "pw30"  "pw31"  "pw32"
## [37] "pw33"  "pw34"  "pw35"  "pw36"  "pw37"  "pw38"  "pw39"  "pw40"  "pw41"
## [46] "pw42"  "pw43"  "pw44"  "pw45"  "pw46"  "pw47"  "pw48"  "pw49"  "pw50"
```

As with other replicate design objects, when printing the object or looking at the summary, the replicate weights are provided along with the data variables.

10.5 Exercises

For this chapter, the exercises entail reading public documentation to determine how to specify the survey design. While reading the documentation, be on the lookout for description of the weights and the survey design variables or replicate weights.

1. The National Health Interview Survey (NHIS) is an annual household survey conducted by the National Center for Health Statistics (NCHS). The NHIS includes a wide variety of health topics for adults including health status and conditions, functioning and disability, health care access and health service utilization, health-related behaviors, health promotion, mental health, barriers to receiving care, and community engagement. Like many national in-person surveys, the sampling design is a stratified clustered design with details included in the Survey Description (National Center for Health Statistics, 2023). The Survey Description provides information on setting up syntax in SUDAAN, Stata, SPSS, SAS, and R ({survey} package implementation). We have imported the data and the variable containing the data as: `nhis_adult_data`. How would we specify the design using either `as_survey_design()` or `as_survey_rep()`?

2. The General Social Survey (GSS) is a survey that has been administered since 1972 on social, behavioral, and attitudinal topics. The 2016-2020 GSS Panel codebook provides examples of setting up syntax in SAS and Stata but not R (Davern et al., 2021). We have imported the data and the variable containing the data as: `gss_data`. How would we specify the design in R using either `as_survey_design()` or `as_survey_rep()`?

11

Missing data

Prerequisites

For this chapter, load the following packages:

```r
library(tidyverse)
library(survey)
library(srvyr)
library(srvyrexploR)
library(naniar)
library(haven)
library(gt)
```

We are using data from ANES and RECS described in Chapter 4. As a reminder, here is the code to create the design objects for each to use throughout this chapter. For ANES, we need to adjust the weight so it sums to the population instead of the sample (see the ANES documentation and Chapter 4 for more information).

```r
targetpop <- 231592693

anes_adjwgt <- anes_2020 %>%
  mutate(Weight = Weight / sum(Weight) * targetpop)

anes_des <- anes_adjwgt %>%
  as_survey_design(
    weights = Weight,
    strata = Stratum,
    ids = VarUnit,
    nest = TRUE
  )
```

For RECS, details are included in the RECS documentation and Chapter 10.

```
recs_des <- recs_2020 %>%
  as_survey_rep(
    weights = NWEIGHT,
    repweights = NWEIGHT1:NWEIGHT60,
    type = "JK1",
    scale = 59 / 60,
    mse = TRUE
  )
```

11.1 Introduction

Missing data in surveys refer to situations where participants do not provide complete responses to survey questions. Respondents may not have seen a question by design. Or, they may not respond to a question for various other reasons, such as not wanting to answer a particular question, not understanding the question, or simply forgetting to answer. Missing data are important to consider and account for, as they can introduce bias and reduce the representativeness of the data. This chapter provides an overview of the types of missing data, how to assess missing data in surveys, and how to conduct analysis when missing data are present. Understanding this complex topic can help ensure accurate reporting of survey results and provide insight into potential changes to the survey design for the future.

11.2 Missing data mechanisms

There are two main categories that missing data typically fall into: missing by design and unintentional missing data. Missing by design is part of the survey plan and can be more easily incorporated into weights and analyses. Unintentional missing data, on the other hand, can lead to bias in survey estimates if not correctly accounted for. Below we provide more information on the types of missing data.

1. Missing by design/questionnaire skip logic: This type of missingness occurs when certain respondents are intentionally directed to skip specific questions based on their previous responses or characteristics. For example, in a survey about employment, if a respondent indicates that they are not employed, they may be directed to skip

questions related to their job responsibilities. Additionally, some surveys randomize questions or modules so that not all participants respond to all questions. In these instances, respondents would have missing data for the modules not randomly assigned to them.

2. Unintentional missing data: This type of missingness occurs when researchers do not intend for there to be missing data on a particular question, for example, if respondents did not finish the survey or refused to answer individual questions. There are three main types of unintentional missing data that each should be considered and handled differently (Mack et al., 2018; Schafer and Graham, 2002):

   a. Missing completely at random (MCAR): The missing data are unrelated to both observed and unobserved data, and the probability of being missing is the same across all cases. For example, if a respondent missed a question because they had to leave the survey early due to an emergency.

   b. Missing at random (MAR): The missing data are related to observed data but not unobserved data, and the probability of being missing is the same within groups. For example, we know the respondents' ages and older respondents choose not to answer specific questions but younger respondents do answer them.

   c. Missing not at random (MNAR): The missing data are related to unobserved data, and the probability of being missing varies for reasons we are not measuring. For example, if respondents with depression do not answer a question about depression severity.

11.3 Assessing missing data

Before beginning an analysis, we should explore the data to determine if there is missing data and what types of missing data are present. Conducting descriptive analysis can help with the analysis and reporting of survey data and can inform the survey design in future studies. For example, large amounts of unexpected missing data may indicate the questions were unclear or difficult to recall. There are several ways to explore missing data, which we walk through below. When assessing the missing data, we recommend using a data.frame object and not the survey object, as most of the analysis is about patterns of records, and weights are not necessary.

11.3.1   Summarize data

A very rudimentary first exploration is to use the summary() function to
summarize the data, which illuminates NA values in the data. Let's look at a
few analytic variables on the ANES 2020 data using summary():

```
anes_2020 %>%
  select(V202051:EarlyVote2020) %>%
  summary()
```

```
##     V202051                  Income7                      Income
##   Min.   :-9.000    $125k or more:1468    Under $9,999     : 647
##   1st Qu.:-1.000    Under $20k   :1076    $50,000-59,999   : 485
##   Median :-1.000    $20k to < 40k:1051    $100,000-109,999: 451
##   Mean   :-0.726    $40k to < 60k: 984    $250,000 or more: 405
##   3rd Qu.:-1.000    $60k to < 80k: 920    $80,000-89,999   : 383
##   Max.   : 3.000    (Other)      :1437    (Other)         :4565
##                     NA's         : 517    NA's            : 517
##     V201617x        V201616        V201615        V201613        V201611
##   Min.   :-9.0    Min.   :-3    Min.   :-3    Min.   :-3    Min.   :-3
##   1st Qu.: 4.0    1st Qu.:-3    1st Qu.:-3    1st Qu.:-3    1st Qu.:-3
##   Median :11.0    Median :-3    Median :-3    Median :-3    Median :-3
##   Mean   :10.4    Mean   :-3    Mean   :-3    Mean   :-3    Mean   :-3
##   3rd Qu.:17.0    3rd Qu.:-3    3rd Qu.:-3    3rd Qu.:-3    3rd Qu.:-3
##   Max.   :22.0    Max.   :-3    Max.   :-3    Max.   :-3    Max.   :-3
##
##     V201610        V201607        Gender           V201600
##   Min.   :-3    Min.   :-3    Male  :3375    Min.   :-9.00
##   1st Qu.:-3    1st Qu.:-3    Female:4027    1st Qu.: 1.00
##   Median :-3    Median :-3    NA's  :  51    Median : 2.00
##   Mean   :-3    Mean   :-3                   Mean   : 1.47
##   3rd Qu.:-3    3rd Qu.:-3                   3rd Qu.: 2.00
##   Max.   :-3    Max.   :-3                   Max.   : 2.00
##
##                    RaceEth          V201549x        V201547z       V201547e
##   White             :5420    Min.   :-9.0    Min.   :-3    Min.   :-3
##   Black             : 650    1st Qu.: 1.0    1st Qu.:-3    1st Qu.:-3
##   Hispanic          : 662    Median : 1.0    Median :-3    Median :-3
##   Asian, NH/PI      : 248    Mean   : 1.5    Mean   :-3    Mean   :-3
##   AI/AN             : 155    3rd Qu.: 2.0    3rd Qu.:-3    3rd Qu.:-3
##   Other/multiple race: 237   Max.   : 6.0    Max.   :-3    Max.   :-3
##   NA's              :  81
##     V201547d        V201547c       V201547b       V201547a        V201546
##   Min.   :-3    Min.   :-3    Min.   :-3    Min.   :-3    Min.   :-9.00
##   1st Qu.:-3    1st Qu.:-3    1st Qu.:-3    1st Qu.:-3    1st Qu.: 2.00
```

```
##   Median :-3    Median :-3    Median :-3    Median :-3    Median : 2.00
##   Mean   :-3    Mean   :-3    Mean   :-3    Mean   :-3    Mean   : 1.84
##   3rd Qu.:-3    3rd Qu.:-3    3rd Qu.:-3    3rd Qu.:-3    3rd Qu.: 2.00
##   Max.   :-3    Max.   :-3    Max.   :-3    Max.   :-3    Max.   : 2.00
##
##        Education          V201510                 AgeGroup
##   Less than HS: 312   Min.   :-9.00    18-29      : 871
##   High school :1160   1st Qu.: 3.00    30-39      :1241
##   Post HS     :2514   Median : 5.00    40-49      :1081
##   Bachelor's  :1877   Mean   : 5.62    50-59      :1200
##   Graduate    :1474   3rd Qu.: 6.00    60-69      :1436
##   NA's        : 116   Max.   :95.00    70 or older:1330
##                                        NA's       : 294
##      Age            V201507x                 TrustPeople
##   Min.   :18.0   Min.   :-9.0   Always            :  48
##   1st Qu.:37.0   1st Qu.:35.0   Most of the time  :3511
##   Median :53.0   Median :51.0   About half the time:2020
##   Mean   :51.8   Mean   :49.4   Some of the time  :1597
##   3rd Qu.:66.0   3rd Qu.:66.0   Never             : 264
##   Max.   :80.0   Max.   :80.0   NA's              :  13
##   NA's   :294
##      V201237              TrustGovernment      V201233
##   Min.   :-9.00   Always            :  80   Min.   :-9.00
##   1st Qu.: 2.00   Most of the time  :1016   1st Qu.: 3.00
##   Median : 3.00   About half the time:2313  Median : 4.00
##   Mean   : 2.78   Some of the time  :3313   Mean   : 3.43
##   3rd Qu.: 3.00   Never             : 702   3rd Qu.: 4.00
##   Max.   : 5.00   NA's              :  29   Max.   : 5.00
##
##              PartyID        V201231x            V201230
##   Strong democrat     :1796  Min.   :-9.00   Min.   :-9.000
##   Strong republican   :1545  1st Qu.: 2.00   1st Qu.:-1.000
##   Independent-democrat: 881  Median : 4.00   Median :-1.000
##   Independent         : 876  Mean   : 3.83   Mean   : 0.013
##   Not very strong democrat: 790  3rd Qu.: 6.00   3rd Qu.: 1.000
##   (Other)             :1540  Max.   : 7.00   Max.   : 3.000
##   NA's                :  25
##      V201229          V201228      VotedPres2016_selection
##   Min.   :-9.000   Min.   :-9.00   Clinton:2911
##   1st Qu.:-1.000   1st Qu.: 1.00   Trump  :2466
##   Median : 1.000   Median : 2.00   Other  : 390
##   Mean   : 0.515   Mean   : 1.99   NA's   :1686
##   3rd Qu.: 1.000   3rd Qu.: 3.00
##   Max.   : 2.000   Max.   : 5.00
##
```

```
##        V201103           VotedPres2016    V201102              V201101
##   Min.     :-9.00      Yes :5810      Min.      :-9.000   Min.     :-9.000
##   1st Qu.:  1.00      No  :1622      1st Qu.:-1.000      1st Qu.:-1.000
##   Median :  1.00      NA's:  21      Median :  1.000     Median :-1.000
##   Mean    :  1.04                     Mean     :  0.105   Mean     :  0.085
##   3rd Qu.:  2.00                     3rd Qu.:  1.000     3rd Qu.:  1.000
##   Max.     :  5.00                     Max.      :  2.000   Max.      :  2.000
##
##        V201029           V201028         V201025x            V201024
##   Min.     :-9.000     Min.     :-9.0    Min.     :-4.00    Min.     :-9.00
##   1st Qu.:-1.000     1st Qu.:-1.0    1st Qu.:  3.00    1st Qu.:-1.00
##   Median :-1.000     Median :-1.0    Median :  3.00    Median :-1.00
##   Mean    :-0.897     Mean     :-0.9    Mean     :  2.92    Mean     :-0.86
##   3rd Qu.:-1.000     3rd Qu.:-1.0    3rd Qu.:  3.00    3rd Qu.:-1.00
##   Max.     :12.000     Max.     :  2.0    Max.     :  4.00    Max.     :  4.00
##
##   EarlyVote2020
##   Yes :  375
##   No  :  115
##   NA's:6963
##
##
##
##
```

We see that there are NA values in several of the derived variables (those not beginning with "V") and negative values in the original variables (those beginning with "V"). We can also use the count() function to get an understanding of the different types of missing data on the original variables. For example, let's look at the count of data for V202072, which corresponds to our VotedPres2020 variable.

```
anes_2020 %>%
  count(VotedPres2020, V202072)
```

```
## # A tibble: 7 x 3
##    VotedPres2020 V202072                                         n
##    <fct>         <dbl+lbl>                                   <int>
## 1 Yes           -1 [-1. Inapplicable]                         361
## 2 Yes            1 [1. Yes, voted for President]             5952
## 3 No            -1 [-1. Inapplicable]                          10
## 4 No             2 [2. No, didn't vote for President]          77
## 5 <NA>          -9 [-9. Refused]                                2
## 6 <NA>          -6 [-6. No post-election interview]             4
## 7 <NA>          -1 [-1. Inapplicable]                         1047
```

Here, we can see that there are three types of missing data, and the majority of them fall under the "Inapplicable" category. This is usually a term associated with data missing due to skip patterns and is considered to be missing data by design. Based on the documentation from ANES (DeBell, 2010), we can see that this question was only asked to respondents who voted in the election.

11.3.2 Visualization of missing data

It can be challenging to look at tables for every variable and instead may be more efficient to view missing data in a graphical format to help narrow in on patterns or unique variables. The {naniar} package is very useful in exploring missing data visually. We can use the vis_miss() function available in both {visdat} and {naniar} packages to view the amount of missing data by variable (see Figure 11.1) (Tierney, 2017; Tierney and Cook, 2023).

```
anes_2020_derived <- anes_2020 %>%
  select(
    -starts_with("V2"), -CaseID, -InterviewMode,
    -Weight, -Stratum, -VarUnit
  )

anes_2020_derived %>%
  vis_miss(cluster = TRUE, show_perc = FALSE) +
  scale_fill_manual(
    values = book_colors[c(3, 1)],
    labels = c("Present", "Missing"),
    name = ""
  ) +
  theme(
    plot.margin = margin(5.5, 30, 5.5, 5.5, "pt"),
    axis.text.x = element_text(angle = 70)
  )
```

From the visualization in Figure 11.1, we can start to get a picture of what questions may be connected in terms of missing data. Even if we did not have the informative variable names, we could deduce that VotedPres2020, VotedPres2020_selection, and EarlyVote2020 are likely connected since their missing data patterns are similar.

Additionally, we can also look at VotedPres2016_selection and see that there are a lot of missing data in that variable. The missing data are likely due to a skip pattern, and we can look at other graphics to see how they relate to other variables. The {naniar} package has multiple visualization functions that can help dive deeper, such as the gg_miss_fct() function, which looks at missing data for all variables by levels of another variable (see Figure 11.2).

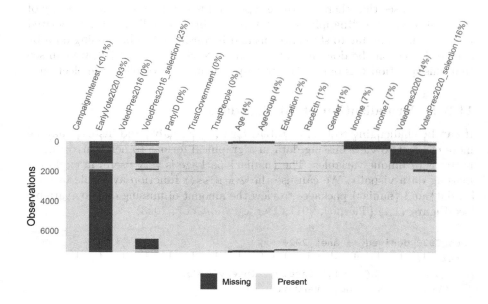

FIGURE 11.1 Visual depiction of missing data in the ANES 2020 data

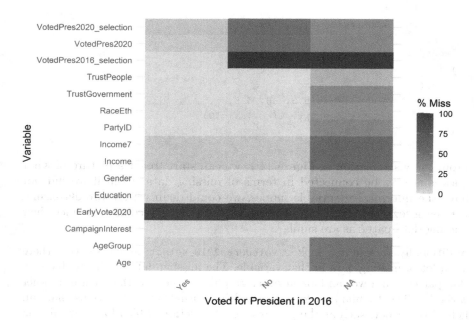

FIGURE 11.2 Missingness in variables for each level of 'VotedPres2016,' in the ANES 2020 data

```
anes_2020_derived %>%
  gg_miss_fct(VotedPres2016) +
  scale_fill_gradientn(
    guide = "colorbar",
    name = "% Miss",
    colors = book_colors[c(3, 2, 1)]
  ) +
  ylab("Variable") +
  xlab("Voted for President in 2016")
```

In Figure 11.2, we can see that if respondents did not vote for president in 2016 or did not answer that question, then they were not asked about who they voted for in 2016 (the percentage of missing data is 100%). Additionally, we can see with Figure 11.2 that there are more missing data across all questions if they did not provide an answer to VotedPres2016.

There are other visualizations that work well with numeric data. For example, in the RECS 2020 data, we can plot two continuous variables and the missing data associated with them to see if there are any patterns in the missingness. To do this, we can use the bind_shadow() function from the {naniar} package. This creates a nabular (combination of "na" with "tabular"), which features the original columns followed by the same number of columns with a specific NA format. These NA columns are indicators of whether the value in the original data is missing or not. The example printed below shows how most levels of HeatingBehavior are not missing (!NA) in the NA variable of HeatingBehavior_NA, but those missing in HeatingBehavior are also missing in HeatingBehavior_NA.

```
recs_2020_shadow <- recs_2020 %>%
  bind_shadow()

ncol(recs_2020)
```

```
## [1] 100
```

```
ncol(recs_2020_shadow)
```

```
## [1] 200
```

```
recs_2020_shadow %>%
  count(HeatingBehavior, HeatingBehavior_NA)
```

```
## # A tibble: 7 x 3
##    HeatingBehavior                         HeatingBehavior_NA       n
##    <fct>                                   <fct>               <int>
## 1 Set one temp and leave it                !NA                 7806
## 2 Manually adjust at night/no one home     !NA                 4654
## 3 Programmable or smart thermostat automatic~ !NA              3310
## 4 Turn on or off as needed                 !NA                 1491
## 5 No control                               !NA                  438
## 6 Other                                    !NA                   46
## 7 <NA>                                     NA                   751
```

We can then use these new variables to plot the missing data alongside the actual data. For example, let's plot a histogram of the total electric bill grouped by those missing and not missing by heating behavior (see Figure 11.3).

```
recs_2020_shadow %>%
  filter(TOTALDOL < 5000) %>%
  ggplot(aes(x = TOTALDOL, fill = HeatingBehavior_NA)) +
  geom_histogram() +
  scale_fill_manual(
    values = book_colors[c(3, 1)],
    labels = c("Present", "Missing"),
    name = "Heating Behavior"
  ) +
  theme_minimal() +
  xlab("Total Energy Cost (Truncated at $5000)") +
  ylab("Number of Households")
```

```
## `stat_bin()` using `bins = 30`. Pick better value with `binwidth`.
```

Figure 11.3 indicates that respondents who did not provide a response for the heating behavior question may have a different distribution of total energy cost compared to respondents who did provide a response. This view of the raw data and missingness could indicate some bias in the data. Researchers take these different bias aspects into account when calculating weights, and we need to make sure that we incorporate the weights when analyzing the data.

There are many other visualizations that can be helpful in reviewing the data, and we recommend reviewing the {naniar} documentation for more information (Tierney and Cook, 2023).

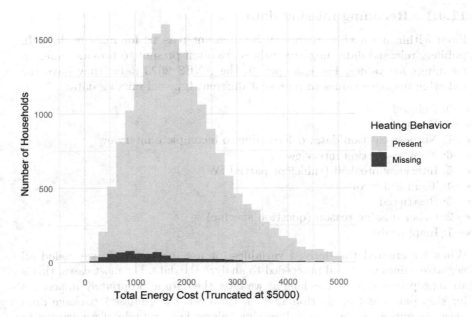

FIGURE 11.3 Histogram of energy cost by heating behavior missing data

11.4 Analysis with missing data

Once we understand the types of missingness, we can begin the analysis of the data. Different missingness types may be handled in different ways. In most publicly available datasets, researchers have already calculated weights and imputed missing values if necessary. Often, there are imputation flags included in the data that indicate if each value in a given variable is imputed. For example, in the RECS data we may see a logical variable of ZWinterTempNight, where a value of TRUE means that the value of WinterTempNight for that respondent was imputed, and FALSE means that it was not imputed. We may use these imputation flags if we are interested in examining the nonresponse rates in the original data. For those interested in learning more about how to calculate weights and impute data for different missing data mechanisms, we recommend Kim and Shao (2021) and Valliant and Dever (2018).

Even with weights and imputation, missing data are most likely still present and need to be accounted for in analysis. This section provides an overview on how to recode missing data in R, and how to account for skip patterns in analysis.

11.4.1   Recoding missing data

Even within a variable, there can be different reasons for missing data. In publicly released data, negative values are often present to provide different meanings for values. For example, in the ANES 2020 data, they have the following negative values to represent different types of missing data:

- −9: Refused
- −8: Don't Know
- −7: No post-election data, deleted due to incomplete interview
- −6: No post-election interview
- −5: Interview breakoff (sufficient partial IW)
- −4: Technical error
- −3: Restricted
- −2: Other missing reason (question specific)
- −1: Inapplicable

When we created the derived variables for use in this book, we coded all negative values as NA and proceeded to analyze the data. For most cases, this is an appropriate approach as long as we filter the data appropriately to account for skip patterns (see Section 11.4.2). However, the {naniar} package does have the option to code special missing values. For example, if we wanted to have two NA values, one that indicated the question was missing by design (e.g., due to skip patterns) and one for the other missing categories, we can use the nabular format to incorporate these with the recode_shadow() function.

```
anes_2020_shadow <- anes_2020 %>%
  select(starts_with("V2")) %>%
  mutate(across(everything(), ~ case_when(
    .x < -1 ~ NA,
    TRUE ~ .x
  ))) %>%
  bind_shadow() %>%
  recode_shadow(V201103 = .where(V201103 == -1 ~ "skip"))

anes_2020_shadow %>%
  count(V201103, V201103_NA)
```

```
## # A tibble: 5 x 3
##    V201103                 V201103_NA      n
##    <dbl+lbl>               <fct>       <int>
## 1 -1 [-1. Inapplicable]    NA_skip      1643
## 2  1 [1. Hillary Clinton]  !NA          2911
## 3  2 [2. Donald Trump]     !NA          2466
## 4  5 [5. Other {SPECIFY}]  !NA           390
## 5 NA                       NA             43
```

However, it is important to note that at the time of publication, there is no easy way to implement recode_shadow() to multiple variables at once (e.g., we cannot use the tidyverse feature of across()). The example code above only implements this for a single variable, so this would have to be done manually or in a loop for all variables of interest.

11.4.2 Accounting for skip patterns

When questions are skipped by design in a survey, it is meaningful that the data are later missing. For example, the RECS asks people how they control the heat in their home in the winter (HeatingBehavior). This is only among those who have heat in their home (SpaceHeatingUsed). If there is no heating equipment used, the value of HeatingBehavior is missing. One has several choices when analyzing these data which include: (1) only including those with a valid value of HeatingBehavior and specifying the universe as those with heat or (2) including those who do not have heat. It is important to specify what population an analysis generalizes to.

Here is an example where we only include those with a valid value of HeatingBehavior (choice 1). Note that we use the design object (recs_des) and then filter to those that are not missing on HeatingBehavior.

```
heat_cntl_1 <- recs_des %>%
  filter(!is.na(HeatingBehavior)) %>%
  group_by(HeatingBehavior) %>%
  summarize(
    p = survey_prop()
  )

heat_cntl_1
```

```
## # A tibble: 6 x 3
##    HeatingBehavior                                            p   p_se
##    <fct>                                                  <dbl>  <dbl>
## 1 Set one temp and leave it                              0.430 4.69e-3
## 2 Manually adjust at night/no one home                   0.264 4.54e-3
## 3 Programmable or smart thermostat automatically adju~   0.168 3.12e-3
## 4 Turn on or off as needed                               0.102 2.89e-3
## 5 No control                                            0.0333 1.70e-3
## 6 Other                                                0.00208 3.59e-4
```

Here is an example where we include those who do not have heat (choice 2). To help understand what we are looking at, we have included the output to show both variables, SpaceHeatingUsed and HeatingBehavior.

```
heat_cntl_2 <- recs_des %>%
  group_by(interact(SpaceHeatingUsed, HeatingBehavior)) %>%
  summarize(
    p = survey_prop()
  )

heat_cntl_2
```

```
## # A tibble: 7 x 4
##   SpaceHeatingUsed HeatingBehavior                          p    p_se
##   <lgl>            <fct>                                <dbl>   <dbl>
## 1 FALSE            <NA>                                0.0469 2.07e-3
## 2 TRUE             Set one temp and leave it           0.410  4.60e-3
## 3 TRUE             Manually adjust at night/no one ho~ 0.251  4.36e-3
## 4 TRUE             Programmable or smart thermostat a~ 0.160  2.95e-3
## 5 TRUE             Turn on or off as needed            0.0976 2.79e-3
## 6 TRUE             No control                          0.0317 1.62e-3
## 7 TRUE             Other                               0.00198 3.41e-4
```

If we ran the first analysis, we would say that 16.8% of households with heat use a programmable or smart thermostat for heating their home. If we used the results from the second analysis, we would say that 16% of households use a programmable or smart thermostat for heating their home. The distinction between the two statements is made bold for emphasis. Skip patterns often change the universe we are talking about and need to be carefully examined.

Filtering to the correct universe is important when handling these types of missing data. The nabular we created above can also help with this. If we have NA_skip values in the shadow, we can make sure that we filter out all of these values and only include relevant missing values. To do this with survey data, we could first create the nabular, then create the design object on that data, and then use the shadow variables to assist with filtering the data. Let's use the nabular we created above for ANES 2020 (anes_2020_shadow) to create the design object.

```
anes_adjwgt_shadow <- anes_2020_shadow %>%
  mutate(V200010b = V200010b / sum(V200010b) * targetpop)

anes_des_shadow <- anes_adjwgt_shadow %>%
  as_survey_design(
    weights = V200010b,
    strata = V200010d,
    ids = V200010c,
    nest = TRUE
  )
```

Then, we can use this design object to look at the percentage of the population who voted for each candidate in 2016 (V201103). First, let's look at the percentages without removing any cases:

```
pres16_select1 <- anes_des_shadow %>%
  group_by(V201103) %>%
  summarize(
    All_Missing = survey_prop()
  )

pres16_select1
```

```
## # A tibble: 5 x 3
##    V201103                    All_Missing All_Missing_se
##    <dbl+lbl>                        <dbl>          <dbl>
## 1 -1 [-1. Inapplicable]            0.324        0.00933
## 2  1 [1. Hillary Clinton]          0.330        0.00728
## 3  2 [2. Donald Trump]             0.299        0.00728
## 4  5 [5. Other {SPECIFY}]          0.0409       0.00230
## 5 NA                               0.00627      0.00121
```

Next, we look at the percentages, removing only those missing due to skip patterns (i.e., they did not receive this question).

```
pres16_select2 <- anes_des_shadow %>%
  filter(V201103_NA != "NA_skip") %>%
  group_by(V201103) %>%
  summarize(
    No_Skip_Missing = survey_prop()
  )

pres16_select2
```

```
## # A tibble: 4 x 3
##   V201103                   No_Skip_Missing No_Skip_Missing_se
##   <dbl+lbl>                           <dbl>              <dbl>
## 1  1 [1. Hillary Clinton]             0.488            0.00870
## 2  2 [2. Donald Trump]               0.443            0.00856
## 3  5 [5. Other {SPECIFY}]            0.0606           0.00330
## 4 NA                                 0.00928          0.00178
```

Finally, we look at the percentages, removing all missing values both due to skip patterns and due to those who refused to answer the question.

```
pres16_select3 <- anes_des_shadow %>%
  filter(V201103_NA == "!NA") %>%
  group_by(V201103) %>%
  summarize(
    No_Missing = survey_prop()
  )


pres16_select3
```

```
## # A tibble: 3 x 3
##    V201103                   No_Missing No_Missing_se
##    <dbl+lbl>                      <dbl>         <dbl>
## 1 1 [1. Hillary Clinton]         0.492       0.00875
## 2 2 [2. Donald Trump]            0.447       0.00861
## 3 5 [5. Other {SPECIFY}]         0.0611      0.00332
```

TABLE 11.1 Percentage of votes by candidate for different missing data inclusions

Candidate	Including All Missing Data		Removing Skip Patterns Only		Removing All Missing Data	
	%	s.e. (%)	%	s.e. (%)	%	s.e. (%)
Did Not Vote for President in 2016	32.4	0.9	NA	NA	NA	NA
Hillary Clinton	33.0	0.7	48.8	0.9	49.2	0.9
Donald Trump	29.9	0.7	44.3	0.9	44.7	0.9
Other Candidate	4.1	0.2	6.1	0.3	6.1	0.3
Missing	0.6	0.1	0.9	0.2	NA	NA

As Table 11.1 shows, the results can vary greatly depending on which type of missing data are removed. If we remove only the skip patterns, the margin between Clinton and Trump is 4.5 percentage points; but if we include all data, even those who did not vote in 2016, the margin is 3.1 percentage points. How we handle the different types of missing values is important for interpreting the data.

12

Successful survey analysis recommendations

Prerequisites

For this chapter, load the following packages:

```
library(tidyverse)
library(survey)
library(srvyr)
library(srvyrexploR)
```

To illustrate the importance of data visualization, we discuss Anscombe's Quartet. The dataset can be replicated by running the code below:

```
anscombe_tidy <- anscombe %>%
  mutate(obs = row_number()) %>%
  pivot_longer(-obs, names_to = "key", values_to = "value") %>%
  separate(key, c("variable", "set"), 1, convert = TRUE) %>%
  mutate(set = c("I", "II", "III", "IV")[set]) %>%
  pivot_wider(names_from = variable, values_from = value)
```

We create an example survey dataset to explain potential pitfalls and how to overcome them in survey analysis. To recreate the dataset, run the code below:

```
example_srvy <- tribble(
  ~id, ~region, ~q_d1, ~q_d2_1, ~gender, ~weight,
  1L, 1L, 1L, "Somewhat interested", "female", 1740,
  2L, 1L, 1L, "Not at all interested", "female", 1428,
  3L, 2L, NA, "Somewhat interested", "female", 496,
  4L, 2L, 1L, "Not at all interested", "female", 550,
  5L, 3L, 1L, "Somewhat interested", "female", 1762,
  6L, 4L, NA, "Very interested", "female", 1004,
  7L, 4L, NA, "Somewhat interested", "female", 522,
  8L, 3L, 2L, "Not at all interested", "female", 1099,
  9L, 4L, 2L, "Somewhat interested", "female", 1295,
  10L, 2L, 2L, "Somewhat interested", "male", 983
)

example_des <-
  example_srvy %>%
  as_survey_design(weights = weight)
```

12.1 Introduction

The previous chapters in this book aimed to provide the technical skills and knowledge required for running survey analyses. This chapter builds upon the previously mentioned best practices to present a curated set of recommendations for running a successful survey analysis. We hope this list provides practical insights that assist in producing meaningful and reliable results.

12.2 Follow the survey analysis process

As we first introduced in Chapter 4, there are four main steps to successfully analyze survey data:

1. Create a `tbl_svy` object (a survey object) using: `as_survey_design()` or `as_survey_rep()`

2. Subset data (if needed) using `filter()` (to create subpopulations)

3. Specify domains of analysis using `group_by()`

4. Within `summarize()`, specify variables to calculate, including means, totals, proportions, quantiles, and more

The order of these steps matters in survey analysis. For example, if we need to subset the data, we must use `filter()` on our data after creating the survey design. If we do this before the survey design is created, we may not be correctly accounting for the study design, resulting in inaccurate findings.

Additionally, correctly identifying the survey design is one of the most important steps in survey analysis. Knowing the type of sample design (e.g., clustered, stratified) helps ensure the underlying error structure is correctly calculated and weights are correctly used. Learning about complex design factors such as clustering, stratification, and weighting is foundational to complex survey analysis, and we recommend that all analysts review Chapter 10 before creating their first design object. Reviewing the documentation (see Chapter 3) helps us understand what variables to use from the data.

Making sure to use the survey analysis functions from the {srvyr} and {survey} packages is also important in survey analysis. For example, using `mean()` and `survey_mean()` on the same data results in different findings and outputs. Each of the survey functions from {srvyr} and {survey} impacts standard errors and variance, and we cannot treat complex surveys as unweighted simple random samples if we want to produce unbiased estimates (Freedman Ellis and Schneider, 2024; Lumley, 2010).

12.3 Begin with descriptive analysis

When receiving a fresh batch of data, it is tempting to jump right into running models to find significant results. However, a successful data analyst begins by exploring the dataset. Chapter 11 talks about the importance of reviewing data when examining missing data patterns. In this chapter, we illustrate the value of reviewing all types of data. This involves running descriptive analysis on the dataset as a whole, as well as individual variables and combinations of variables. As described in Chapter 5, descriptive analyses should always precede statistical analysis to prevent avoidable (and potentially embarrassing) mistakes.

12.3.1 Table review

Even before applying weights, consider running cross-tabulations on the raw data. Cross-tabs can help us see if any patterns stand out that may be alarming or something worth further investigating.

For example, let's explore the example survey dataset introduced in the Prerequisites box, `example_srvy`. We run the code below on the unweighted data to inspect the `gender` variable:

```
example_srvy %>%
  group_by(gender) %>%
  summarize(n = n())
```

```
## # A tibble: 2 x 2
##   gender     n
##   <chr>  <int>
## 1 female     9
## 2 male       1
```

The data show that females comprise 9 out of 10, or 90%, of the sample. Generally, we assume something close to a 50/50 split between male and female respondents in a population. The sizable female proportion could indicate either a unique sample or a potential error in the data. If we review the survey documentation and see this was a deliberate part of the design, we can continue our analysis using the appropriate methods. If this was not an intentional choice by the researchers, the results alert us that something may be incorrect in the data or our code, and we can verify if there's an issue by comparing the results with the weighted means.

12.3.2 Graphical review

Tables provide a quick check of our assumptions, but there is no substitute for graphs and plots to visualize the distribution of data. We might miss outliers or nuances if we scan only summary statistics.

For example, Anscombe's Quartet demonstrates the importance of visualization in analysis. Let's say we have a dataset with x- and y-variables in an object called `anscombe_tidy`. Let's take a look at how the dataset is structured:

```
head(anscombe_tidy)
```

```
## # A tibble: 6 x 4
##     obs set       x     y
##   <int> <chr> <dbl> <dbl>
## 1     1 I        10  8.04
```

```
## 2      1 II       10  9.14
## 3      1 III      10  7.46
## 4      1 IV        8  6.58
## 5      2 I         8  6.95
## 6      2 II        8  8.14
```

We can begin by checking one set of variables. For Set I, the x-variables have an average of 9 with a standard deviation of 3.3; for y, we have an average of 7.5 with a standard deviation of 2.03. The two variables have a correlation of 0.81.

```
anscombe_tidy %>%
  filter(set == "I") %>%
  summarize(
    x_mean = mean(x),
    x_sd = sd(x),
    y_mean = mean(y),
    y_sd = sd(y),
    correlation = cor(x, y)
  )
```

```
## # A tibble: 1 x 5
##    x_mean  x_sd y_mean  y_sd correlation
##     <dbl> <dbl>  <dbl> <dbl>       <dbl>
## 1       9  3.32   7.50  2.03       0.816
```

These are useful statistics. We can note that the data do not have high variability, and the two variables are strongly correlated. Now, let's check all the sets (I-IV) in the Anscombe data. Notice anything interesting?

```
anscombe_tidy %>%
  group_by(set) %>%
  summarize(
    x_mean = mean(x),
    x_sd = sd(x, na.rm = TRUE),
    y_mean = mean(y),
    y_sd = sd(y, na.rm = TRUE),
    correlation = cor(x, y)
  )
```

```
## # A tibble: 4 x 6
##    set   x_mean x_sd y_mean y_sd correlation
##    <chr>  <dbl> <dbl>  <dbl> <dbl>      <dbl>
## 1 I          9  3.32   7.50  2.03      0.816
## 2 II         9  3.32   7.50  2.03      0.816
## 3 III        9  3.32   7.5   2.03      0.816
## 4 IV         9  3.32   7.50  2.03      0.817
```

The summary results for these four sets are nearly identical! Based on this, we might assume that each distribution is similar. Let's look at a graphical visualization to see if our assumption is correct (see Figure 12.1).

```
ggplot(anscombe_tidy, aes(x, y)) +
  geom_point() +
  facet_wrap(~set) +
  geom_smooth(method = "lm", se = FALSE, alpha = 0.5) +
  theme_minimal()
```

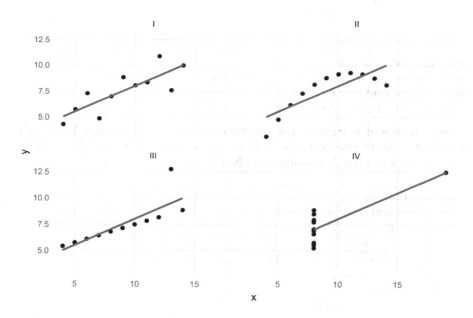

FIGURE 12.1 Plot of Anscombe's Quartet data and the importance of reviewing data graphically

Although each of the four sets has the same summary statistics and regression line, when reviewing the plots (see Figure 12.1), it becomes apparent that the distributions of the data are not the same at all. Each set of points results in different shapes and distributions. Imagine sharing each set (I-IV) and the corresponding plot with a different colleague. The interpretations and descriptions of the data would be very different even though the statistics are similar. Plotting data can also ensure that we are using the correct analysis method on the data, so understanding the underlying distributions is an important first step.

12.4 Check variable types

When we pull the data from surveys into R, the data may be listed as character, factor, numeric, or logical/Boolean. The tidyverse functions that read in data (e.g., `read_csv()`, `read_excel()`) default to have all strings load as character variables. This is important when dealing with survey data, as many strings may be better suited for factors than character variables. For example, let's revisit the `example_srvy` data. Taking a `glimpse()` of the data gives us insight into what it contains:

```
example_srvy %>%
  glimpse()
```

```
## Rows: 10
## Columns: 6
## $ id     <int> 1, 2, 3, 4, 5, 6, 7, 8, 9, 10
## $ region <int> 1, 1, 2, 2, 3, 4, 4, 3, 4, 2
## $ q_d1   <int> 1, 1, NA, 1, 1, NA, NA, 2, 2, 2
## $ q_d2_1 <chr> "Somewhat interested", "Not at all interested", "Some~
## $ gender <chr> "female", "female", "female", "female", "female", "fe~
## $ weight <dbl> 1740, 1428, 496, 550, 1762, 1004, 522, 1099, 1295, 983
```

The output shows that `q_d2_1` is a character variable, but the values of that variable show three options (Very interested / Somewhat interested / Not at all interested). In this case, we most likely want to change `q_d2_1` to be a factor variable and order the factor levels to indicate that this is an ordinal variable. Here is some code on how we might approach this task using the {forcats} package (Wickham, 2023):

```
example_srvy_fct <- example_srvy %>%
  mutate(q_d2_1_fct = factor(
    q_d2_1,
    levels = c(
      "Very interested",
      "Somewhat interested",
      "Not at all interested"
```

```
    )
  ))

example_srvy_fct %>%
  glimpse()
```

```
## Rows: 10
## Columns: 7
## $ id        <int> 1, 2, 3, 4, 5, 6, 7, 8, 9, 10
## $ region    <int> 1, 1, 2, 2, 3, 4, 4, 3, 4, 2
## $ q_d1      <int> 1, 1, NA, 1, 1, NA, NA, 2, 2, 2
## $ q_d2_1    <chr> "Somewhat interested", "Not at all interested", "~
## $ gender    <chr> "female", "female", "female", "female", "female",~
## $ weight    <dbl> 1740, 1428, 496, 550, 1762, 1004, 522, 1099, 1295~
## $ q_d2_1_fct <fct> Somewhat interested, Not at all interested, Somew~
```

```
example_srvy_fct %>%
  count(q_d2_1_fct, q_d2_1)
```

```
## # A tibble: 3 x 3
##   q_d2_1_fct               q_d2_1                     n
##   <fct>                    <chr>                  <int>
## 1 Very interested          Very interested            1
## 2 Somewhat interested      Somewhat interested        6
## 3 Not at all interested    Not at all interested      3
```

This example dataset also includes a column called region, which is imported as a number (<int>). This is a good reminder to use the questionnaire and codebook along with the data to find out if the values actually reflect a number or are perhaps a coded categorical variable (see Chapter 3 for more details). R calculates the mean even if it is not appropriate, leading to the common mistake of applying an average to categorical values instead of a proportion function. For example, for ease of coding, we may use the across() function to calculate the mean across all numeric variables:

```
example_des %>%
  select(-weight) %>%
  summarize(across(where(is.numeric), ~ survey_mean(.x, na.rm = TRUE)))
```

```
## # A tibble: 1 x 6
##      id id_se region region_se  q_d1 q_d1_se
##   <dbl> <dbl>  <dbl>     <dbl> <dbl>   <dbl>
## 1  5.24  1.12   2.49     0.428  1.38   0.196
```

In this example, if we do not adjust `region` to be a factor variable type, we might accidentally report an average region of 2.49 in our findings, which is meaningless. Checking that our variables are appropriate avoids this pitfall and ensures the measures and models are suitable for the variable type.

12.5 Improve debugging skills

It is common for analysts working in R to come across warning or error messages, and learning how to debug these messages (i.e., find and fix issues) ensures we can proceed with our work and avoid potential mistakes.

We've discussed a few examples in this book. For example, if we calculate an average with `survey_mean()` and get NA instead of a number, it may be because our column has missing values.

```
example_des %>%
  summarize(mean = survey_mean(q_d1))
```

```
## # A tibble: 1 x 2
##    mean mean_se
##   <dbl>  <dbl>
## 1    NA    NaN
```

Including the `na.rm = TRUE` would resolve the issue:

```
example_des %>%
  summarize(mean = survey_mean(q_d1, na.rm = TRUE))
```

```
## # A tibble: 1 x 2
##    mean mean_se
##   <dbl>  <dbl>
## 1  1.38   0.196
```

Another common error message that we may see with survey analysis may look something like the following:

```
example_des %>%
  svyttest(q_d1 ~ gender)
```

```
## Error in UseMethod("svymean", design): no applicable method for
##   'svymean' applied to an object of class "formula"
```

In this case, we need to remember that with functions from the {survey} packages like svyttest(), the design object is not the first argument, and we have to use the dot (.) notation (see Chapter 6). Adding in the named argument of design=. fixes this error.

```
example_des %>%
  svyttest(q_d1 ~ gender,
    design = .
  )
```

```
##
##  Design-based t-test
##
## data:  q_d1 ~ gender
## t = 3.5, df = 5, p-value = 0.02
## alternative hypothesis: true difference in mean is not equal to 0
## 95 percent confidence interval:
##   0.1878 1.2041
## sample estimates:
## difference in mean
##               0.696
```

Often, debugging involves interpreting the message from R. For example, if our code results in this error:

```
Error in `contrasts<-`(`*tmp*`, value = contr.funs[1 + isOF[nn]]) :
  contrasts can be applied only to factors with 2 or more levels
```

We can see that the error has to do with a function requiring a factor with two or more levels and that it has been applied to something else. This ties back to our section on using appropriate variable types. We can check the variable of interest to examine whether it is the correct type.

The internet also offers many resources for debugging. Searching for a specific error message can often lead to a solution. In addition, we can post on community forums like Posit Community[1] for direct help from others.

12.6   Think critically about conclusions

Once we have our findings, we need to learn to think critically about them. As mentioned in Chapter 2, many aspects of the study design can impact our interpretation of the results, for example, the number and types of response

---

[1]https://forum.posit.co/

options provided to the respondent or who was asked the question (both thinking about the full sample and any skip patterns). Knowing the overall study design can help us accurately think through what the findings may mean and identify any issues with our analyses. Additionally, we should make sure that our survey design object is correctly defined (see Chapter 10), carefully consider how we are managing missing data (see Chapter 11), and follow statistical analysis procedures such as avoiding model overfitting by using too many variables in our formulas.

These considerations allow us to conduct our analyses and review findings for statistically significant results. It is important to note that even significant results do not mean that they are meaningful or important. A large enough sample can produce statistically significant results. Therefore, we want to look at our results in context, such as comparing them with results from other studies or analyzing them in conjunction with confidence intervals and other measures.

Communicating the results (see Chapter 8) in an unbiased manner is also a critical step in any analysis project. If we present results without error measures or only present results that support our initial hypotheses, we are not thinking critically and may incorrectly represent the data. As survey data analysts, we often interpret the survey data for the public. We must ensure that we are the best stewards of the data and work to bring light to meaningful and interesting findings that the public wants and needs to know about.

Part V

Vignettes

13

National Crime Victimization Survey vignette

Prerequisites

For this chapter, load the following packages:

```
library(tidyverse)
library(survey)
library(srvyr)
library(srvyrexploR)
library(gt)
```

We use data from the United States National Crime Victimization Survey (NCVS). These data are available in the {srvyrexploR} package as ncvs_2021_incident, ncvs_2021_household, and ncvs_2021_person.

13.1 Introduction

The National Crime Victimization Survey (NCVS) is a household survey sponsored by the Bureau of Justice Statistics (BJS), which collects data on criminal victimization, including characteristics of the crimes, offenders, and victims. Crime types include both household and personal crimes, as well as violent and non-violent crimes. The population of interest of this survey is all people in the United States age 12 and older living in housing units and non-institutional group quarters.

The NCVS has been ongoing since 1992. An earlier survey, the National Crime Survey, was run from 1972 to 1991 (U. S. Bureau of Justice Statistics, 2017). The survey is administered using a rotating panel. When an address enters the sample, the residents of that address are interviewed every 6 months for a total of 7 interviews. If the initial residents move away from the address during the period and new residents move in, the new residents are included in the survey, as people are not followed when they move.

NCVS data are publicly available and distributed by Inter-university Consortium for Political and Social Research (ICPSR), with data going back to 1992. The vignette in this book includes data from 2021 (U.S. Bureau of Justice Statistics, 2022). The NCVS data structure is complicated, and the User's Guide contains examples for analysis in SAS, SUDAAN, SPSS, and Stata, but not R (Shook-Sa et al., 2015). This vignette adapts those examples for R.

13.2   Data structure

The data from ICPSR are distributed with five files, each having its unique identifier indicated:

- Address Record - YEARQ, IDHH
- Household Record - YEARQ, IDHH
- Person Record - YEARQ, IDHH, IDPER
- Incident Record - YEARQ, IDHH, IDPER
- 2021 Collection Year Incident - YEARQ, IDHH, IDPER

In this vignette, we focus on the household, person, and incident files and have selected a subset of columns for use in the examples. We have included data in the {srvyexploR} package with this subset of columns, but the complete data files can be downloaded from ICPSR[1].

13.3   Survey notation

The NCVS User Guide (Shook-Sa et al., 2015) uses the following notation:

- i represents NCVS households, identified on the household-level file with the household identification number IDHH.
- j represents NCVS individual respondents within household i, identified on the person-level file with the person identification number IDPER.
- k represents reporting periods (i.e., YEARQ) for household i and individual respondent j.
- l represents victimization records for respondent j in household i and reporting period k. Each record on the NCVS incident-level file is associated with a victimization record l.
- D represents one or more domain characteristics of interest in the calculation of NCVS estimates. For victimization totals and proportions, domains can be defined on the basis of crime types (e.g., violent crimes, property crimes),

---

[1]https://www.icpsr.umich.edu/web/NACJD/studies/38429

characteristics of victims (e.g., age, sex, household income), or characteristics of the victimizations (e.g., victimizations reported to police, victimizations committed with a weapon present). Domains could also be a combination of all of these types of characteristics. For example, in the calculation of victimization rates, domains are defined on the basis of the characteristics of the victims.

- A_a represents the level a of covariate A. Covariate A is defined in the calculation of victimization proportions and represents the characteristic we want to obtain the distribution of victimizations in domain D.
- C represents the personal or property crime for which we want to obtain a victimization rate.

In this vignette, we discuss four estimates:

1. Victimization totals estimate the number of criminal victimizations with a given characteristic. As demonstrated below, these can be calculated from any of the data files. The estimated victimization total, $\hat{t}_D$ for domain D is estimated as

$$\hat{t}_D = \sum_{ijkl \in D} v_{ijkl}$$

where v_{ijkl} is the series-adjusted victimization weight for household i, respondent j, reporting period k, and victimization l, represented in the data as WGTVICCY.

2. Victimization proportions estimate characteristics among victimizations or victims. Victimization proportions are calculated using the incident data file. The estimated victimization proportion for domain D across level a of covariate A, $\hat{p}_{A_a,D}$ is

$$\hat{p}_{A_a,D} = \frac{\sum_{ijkl \in A_a,D} v_{ijkl}}{\sum_{ijkl \in D} v_{ijkl}}.$$

The numerator is the number of incidents with a particular characteristic in a domain, and the denominator is the number of incidents in a domain.

3. Victimization rates are estimates of the number of victimizations per 1,000 persons or households in the population[2]. Victimization rates are calculated using the household or person-level data files. The estimated victimization rate for crime C in domain D is

$$\hat{V}R_{C,D} = \frac{\sum_{ijkl \in C,D} v_{ijkl}}{\sum_{ijk \in D} w_{ijk}} \times 1000$$

---

[2]BJS publishes victimization rates per 1,000, which are also presented in these examples.

where w_{ijk} is the person weight (WGTPERCY) for personal crimes or household weight (WGTHHCY) for household crimes. The numerator is the number of incidents in a domain, and the denominator is the number of persons or households in a domain. Notice that the weights in the numerator and denominator are different; this is important, and in the syntax and examples below, we discuss how to make an estimate that involves two weights.

4. Prevalence rates are estimates of the percentage of the population (persons or households) who are victims of a crime. These are estimated using the household or person-level data files. The estimated prevalence rate for crime C in domain D is

$$\hat{PR}_{C,D} = \frac{\sum_{ijk \in C,D} I_{ij} w_{ijk}}{\sum_{ijk \in D} w_{ijk}} \times 100$$

where I_{ij} is an indicator that a person or household in domain D was a victim of crime C at any time in the year. The numerator is the number of victims in domain D for crime C, and the denominator is the number of people or households in the population.

13.4   Data file preparation

Some work is necessary to prepare the files before analysis. The design variables indicating pseudo-stratum (V2117) and half-sample code (V2118) are only included on the household file, so they must be added to the person and incident files for any analysis.

For victimization rates, we need to know the victimization status for both victims and non-victims. Therefore, the incident file must be summarized and merged onto the household or person files for household-level and person-level crimes, respectively. We begin this vignette by discussing how to create these incident summary files. This is following Section 2.2 of the NCVS User's Guide (Shook-Sa et al., 2015).

13.4.1   Preparing files for estimation of victimization rates

Each record on the incident file represents one victimization, which is not the same as one incident. Some victimizations have several instances that make it difficult for the victim to differentiate the details of these incidents, labeled as "series crimes." Appendix A of the User's Guide indicates how to calculate the series weight in other statistical languages.

Here, we adapt that code for R. Essentially, if a victimization is a series crime, its series weight is top-coded at 10 based on the number of actual victimizations, that is, even if the crime occurred more than 10 times, it is counted as 10 times to reduce the influence of extreme outliers. If an incident is a series crime, but the number of occurrences is unknown, the series weight is set to 6. A description of the variables used to create indicators of series and the associated weights is included in Table 13.1.

TABLE 13.1 Codebook for incident variables, related to series weight

	Description	Value	Label
V4016	How many times incident occur last 6 months	1–996	Number of times
		997	Don't know
V4017	How many incidents	1	1–5 incidents (not a "series")
		2	6 or more incidents
		8	Residue (invalid data)
V4018	Incidents similar in detail	1	Similar
		2	Different (not in a "series")
		8	Residue (invalid data)
V4019	Enough detail to distinguish incidents	1	Yes (not a "series")
		2	No (is a "series")
		8	Residue (invalid data)
WGTVICCY	Adjusted victimization weight		Numeric

We want to create four variables to indicate if an incident is a series crime. First, we create a variable called series using V4017, V4018, and V4019 where an incident is considered a series crime if there are 6 or more incidents (V4107), the incidents are similar in detail (V4018), or there is not enough detail to distinguish the incidents (V4019). Second, we top-code the number of incidents (V4016) by creating a variable n10v4016, which is set to 10 if V4016 > 10. Third, we create the serieswgt using the two new variables series and n10v4019 to classify the max series based on missing data and number of incidents. Finally, we create the new weight using our new serieswgt variable and the existing weight (WGTVICCY).

```
inc_series <- ncvs_2021_incident %>%
  mutate(
    series = case_when(
      V4017 %in% c(1, 8) ~ 1,
      V4018 %in% c(2, 8) ~ 1,
      V4019 %in% c(1, 8) ~ 1,
      TRUE ~ 2
    ),
    n10v4016 = case_when(
      V4016 %in% c(997, 998) ~ NA_real_,
      V4016 > 10 ~ 10,
      TRUE ~ V4016
    ),
    serieswgt = case_when(
      series == 2 & is.na(n10v4016) ~ 6,
      series == 2 ~ n10v4016,
      TRUE ~ 1
    ),
    NEWWGT = WGTVICCY * serieswgt
  )
```

The next step in preparing the files for estimation is to create indicators on the victimization file for characteristics of interest. Almost all BJS publications limit the analysis to records where the victimization occurred in the United States (where V4022 is not equal to 1). We do this for all estimates as well. A brief codebook of variables for this task is located in Table 13.2.

TABLE 13.2 Codebook for incident variables, crime type indicators and characteristics

Variable	Description	Value	Label
V4022	In what city/town/village	1	Outside U.S.
		2	Not inside a city/town/village
		3	Same city/town/village as present residence
		4	Different city/town/village as present residence
		5	Don't know
		6	Don't know if 2, 4, or 5
V4049	Did offender have a weapon	1	Yes

Variable	Description	Value	Label
		2	No
		3	Don't know
V4050	What was the weapon that offender had	1	At least one good entry
		3	Indicates "Yes-Type Weapon-NA"
		7	Indicates "Gun Type Unknown"
		8	No good entry
V4051	Hand gun	0	No
		1	Yes
V4052	Other gun	0	No
		1	Yes
V4053	Knife	0	No
		1	Yes
V4399	Reported to police	1	Yes
		2	No
		3	Don't know
V4529	Type of crime code	01	Completed rape
		02	Attempted rape
		03	Sexual attack with serious assault
		04	Sexual attack with minor assault
		05	Completed robbery with injury from serious assault
		06	Completed robbery with injury from minor assault
		07	Completed robbery without injury from minor assault
		08	Attempted robbery with injury from serious assault
		09	Attempted robbery with injury from minor assault
		10	Attempted robbery without injury
		11	Completed aggravated assault with injury
		12	Attempted aggravated assault with weapon

Variable	Description	Value	Label
		13	Threatened assault with weapon
		14	Simple assault completed with injury
		15	Sexual assault without injury
		16	Unwanted sexual contact without force
		17	Assault without weapon without injury
		18	Verbal threat of rape
		19	Verbal threat of sexual assault
		20	Verbal threat of assault
		21	Completed purse snatching
		22	Attempted purse snatching
		23	Pocket picking (completed only)
		31	Completed burglary, forcible entry
		32	Completed burglary, unlawful entry without force
		33	Attempted forcible entry
		40	Completed motor vehicle theft
		41	Attempted motor vehicle theft
		54	Completed theft less than $10
		55	Completed theft $10 to $49
		56	Completed theft $50 to $249
		57	Completed theft $250 or greater
		58	Completed theft value NA
		59	Attempted theft

Using these variables, we create the following indicators:

1. Property crime
    - $V4529 \geq 31$
    - Variable: `Property`
2. Violent crime
    - $V4529 \leq 20$
    - Variable: `Violent`
3. Property crime reported to the police
    - $V4529 \geq 31$ and $V4399=1$
    - Variable: `Property_ReportPolice`
4. Violent crime reported to the police
    - $V4529 < 31$ and $V4399=1$
    - Variable: `Violent_ReportPolice`
5. Aggravated assault without a weapon
    - $V4529$ in $11:12$ and $V4049=2$
    - Variable: `AAST_NoWeap`
6. Aggravated assault with a firearm
    - $V4529$ in $11:12$ and $V4049=1$ and ($V4051=1$ or $V4052=1$ or $V4050=7$)
    - Variable: `AAST_Firearm`
7. Aggravated assault with a knife or sharp object
    - $V4529$ in $11:12$ and $V4049=1$ and ($V4053=1$ or $V4054=1$)
    - Variable: `AAST_Knife`
8. Aggravated assault with another type of weapon
    - $V4529$ in $11:12$ and $V4049=1$ and $V4050=1$ and not firearm or knife
    - Variable: `AAST_Other`

```
inc_ind <- inc_series %>%
  filter(V4022 != 1) %>%
  mutate(
    WeapCat = case_when(
      is.na(V4049) ~ NA_character_,
      V4049 == 2 ~ "NoWeap",
      V4049 == 3 ~ "UnkWeapUse",
      V4050 == 3 ~ "Other",
      V4051 == 1 | V4052 == 1 | V4050 == 7 ~ "Firearm",
      V4053 == 1 | V4054 == 1 ~ "Knife",
      TRUE ~ "Other"
    ),
    V4529_num = parse_number(as.character(V4529)),
    ReportPolice = V4399 == 1,
    Property = V4529_num >= 31,
```

```
    Violent = V4529_num <= 20,
    Property_ReportPolice = Property & ReportPolice,
    Violent_ReportPolice = Violent & ReportPolice,
    AAST = V4529_num %in% 11:13,
    AAST_NoWeap = AAST & WeapCat == "NoWeap",
    AAST_Firearm = AAST & WeapCat == "Firearm",
    AAST_Knife = AAST & WeapCat == "Knife",
    AAST_Other = AAST & WeapCat == "Other"
  )
```

This is a good point to pause to look at the output of crosswalks between an original variable and a derived one to check that the logic was programmed correctly and that everything ends up in the expected category.

```
inc_series %>% count(V4022)
```

```
## # A tibble: 6 x 2
##   V4022      n
##   <fct> <int>
## 1 1        34
## 2 2        65
## 3 3      7697
## 4 4      1143
## 5 5        39
## 6 8         4
```

```
inc_ind %>% count(V4022)
```

```
## # A tibble: 5 x 2
##   V4022      n
##   <fct> <int>
## 1 2        65
## 2 3      7697
## 3 4      1143
## 4 5        39
## 5 8         4
```

```
inc_ind %>%
  count(WeapCat, V4049, V4050, V4051, V4052, V4052, V4053, V4054)
```

```
## # A tibble: 13 x 8
##    WeapCat     V4049 V4050 V4051 V4052 V4053 V4054     n
##    <chr>       <fct> <fct> <fct> <fct> <fct> <fct> <int>
##  1 Firearm     1     1     0     1     0     0        15
##  2 Firearm     1     1     0     1     1     1         1
##  3 Firearm     1     1     1     0     0     0       125
##  4 Firearm     1     1     1     0     1     0         2
##  5 Firearm     1     1     1     1     0     0         3
##  6 Firearm     1     7     0     0     0     0         3
##  7 Knife       1     1     0     0     0     1        14
##  8 Knife       1     1     0     0     1     0        71
##  9 NoWeap      2     <NA>  <NA>  <NA>  <NA>  <NA>   1794
## 10 Other       1     1     0     0     0     0       147
## 11 Other       1     3     0     0     0     0        26
## 12 UnkWeapUse  3     <NA>  <NA>  <NA>  <NA>  <NA>    519
## 13 <NA>        <NA>  <NA>  <NA>  <NA>  <NA>  <NA>   6228
```

```r
inc_ind %>%
  count(V4529, Property, Violent, AAST) %>%
  print(n = 40)
```

```
## # A tibble: 34 x 5
##    V4529 Property Violent AAST      n
##    <fct> <lgl>    <lgl>   <lgl> <int>
##  1 1     FALSE    TRUE    FALSE    45
##  2 2     FALSE    TRUE    FALSE    20
##  3 3     FALSE    TRUE    FALSE    11
##  4 4     FALSE    TRUE    FALSE     3
##  5 5     FALSE    TRUE    FALSE    24
##  6 6     FALSE    TRUE    FALSE    26
##  7 7     FALSE    TRUE    FALSE    59
##  8 8     FALSE    TRUE    FALSE     5
##  9 9     FALSE    TRUE    FALSE     7
## 10 10    FALSE    TRUE    FALSE    57
## 11 11    FALSE    TRUE    TRUE     97
## 12 12    FALSE    TRUE    TRUE     91
## 13 13    FALSE    TRUE    TRUE    163
## 14 14    FALSE    TRUE    FALSE   165
## 15 15    FALSE    TRUE    FALSE    24
## 16 16    FALSE    TRUE    FALSE    12
## 17 17    FALSE    TRUE    FALSE   357
## 18 18    FALSE    TRUE    FALSE    14
## 19 19    FALSE    TRUE    FALSE     3
## 20 20    FALSE    TRUE    FALSE   607
## 21 21    FALSE    FALSE   FALSE     2
```

```
## 22 22    FALSE    FALSE    FALSE      2
## 23 23    FALSE    FALSE    FALSE     19
## 24 31    TRUE     FALSE    FALSE    248
## 25 32    TRUE     FALSE    FALSE    634
## 26 33    TRUE     FALSE    FALSE    188
## 27 40    TRUE     FALSE    FALSE    256
## 28 41    TRUE     FALSE    FALSE     97
## 29 54    TRUE     FALSE    FALSE    407
## 30 55    TRUE     FALSE    FALSE   1006
## 31 56    TRUE     FALSE    FALSE   1686
## 32 57    TRUE     FALSE    FALSE   1420
## 33 58    TRUE     FALSE    FALSE    798
## 34 59    TRUE     FALSE    FALSE    395
```

```
inc_ind %>% count(ReportPolice, V4399)
```

```
## # A tibble: 4 x 3
##    ReportPolice V4399       n
##    <lgl>        <fct>   <int>
## 1 FALSE         2        5670
## 2 FALSE         3         103
## 3 FALSE         8          12
## 4 TRUE          1        3163
```

```
inc_ind %>%
  count(
    AAST,
    WeapCat,
    AAST_NoWeap,
    AAST_Firearm,
    AAST_Knife,
    AAST_Other
  )
```

```
## # A tibble: 11 x 7
##    AAST  WeapCat AAST_NoWeap AAST_Firearm AAST_Knife AAST_Other     n
##    <lgl> <chr>   <lgl>       <lgl>        <lgl>      <lgl>      <int>
## 1 FALSE Firearm  FALSE       FALSE        FALSE      FALSE         34
## 2 FALSE Knife    FALSE       FALSE        FALSE      FALSE         23
## 3 FALSE NoWeap   FALSE       FALSE        FALSE      FALSE       1769
## 4 FALSE Other    FALSE       FALSE        FALSE      FALSE         27
## 5 FALSE UnkWeap~ FALSE       FALSE        FALSE      FALSE        516
## 6 FALSE <NA>     FALSE       FALSE        FALSE      FALSE       6228
## 7 TRUE  Firearm  FALSE       TRUE         FALSE      FALSE        115
## 8 TRUE  Knife    FALSE       FALSE        TRUE       FALSE         62
```

```
##  9 TRUE  NoWeap   TRUE     FALSE     FALSE     FALSE     25
## 10 TRUE  Other    FALSE    FALSE     FALSE     TRUE     146
## 11 TRUE  UnkWeap~ FALSE    FALSE     FALSE     FALSE      3
```

After creating indicators of victimization types and characteristics, the file is summarized, and crimes are summed across persons or households by YEARQ. Property crimes (i.e., crimes committed against households, such as household burglary or motor vehicle theft) are summed across households, and personal crimes (i.e., crimes committed against an individual, such as assault, robbery, and personal theft) are summed across persons. The indicators are summed using our created series weight variable (serieswgt). Additionally, the existing weight variable (WGTVICCY) needs to be retained for later analysis.

```
inc_hh_sums <-
  inc_ind %>%
  filter(V4529_num > 23) %>% # restrict to household crimes
  group_by(YEARQ, IDHH) %>%
  summarize(
    WGTVICCY = WGTVICCY[1],
    across(starts_with("Property"),
      ~ sum(. * serieswgt),
      .names = "{.col}"
    ),
    .groups = "drop"
  )

inc_pers_sums <-
  inc_ind %>%
  filter(V4529_num <= 23) %>% # restrict to person crimes
  group_by(YEARQ, IDHH, IDPER) %>%
  summarize(
    WGTVICCY = WGTVICCY[1],
    across(c(starts_with("Violent"), starts_with("AAST")),
      ~ sum(. * serieswgt),
      .names = "{.col}"
    ),
    .groups = "drop"
  )
```

Now, we merge the victimization summary files into the appropriate files. For any record on the household or person file that is not on the victimization file, the victimization counts are set to 0 after merging. In this step, we also create the victimization adjustment factor. See Section 2.2.4 in the User's Guide for details of why this adjustment is created (Shook-Sa et al., 2015). It is calculated as follows:

$$A_{ijk} = \frac{v_{ijk}}{w_{ijk}}$$

where w_{ijk} is the person weight (WGTPERCY) for personal crimes or the household weight (WGTHHCY) for household crimes, and v_{ijk} is the victimization weight (WGTVICCY) for household i, respondent j, in reporting period k. The adjustment factor is set to 0 if no incidents are reported.

```
hh_z_list <- rep(0, ncol(inc_hh_sums) - 3) %>%
  as.list() %>%
  setNames(names(inc_hh_sums)[-(1:3)])
pers_z_list <- rep(0, ncol(inc_pers_sums) - 4) %>%
  as.list() %>%
  setNames(names(inc_pers_sums)[-(1:4)])

hh_vsum <- ncvs_2021_household %>%
  full_join(inc_hh_sums, by = c("YEARQ", "IDHH")) %>%
  replace_na(hh_z_list) %>%
  mutate(ADJINC_WT = if_else(is.na(WGTVICCY), 0, WGTVICCY / WGTHHCY))

pers_vsum <- ncvs_2021_person %>%
  full_join(inc_pers_sums, by = c("YEARQ", "IDHH", "IDPER")) %>%
  replace_na(pers_z_list) %>%
  mutate(ADJINC_WT = if_else(is.na(WGTVICCY), 0, WGTVICCY / WGTPERCY))
```

13.4.2　Derived demographic variables

A final step in file preparation for the household and person files is creating any derived variables on the household and person files, such as income categories or age categories, for subgroup analysis. We can do this step before or after merging the victimization counts.

13.4.2.1　Household variables

For the household file, we create categories for tenure (rental status), urbanicity, income, place size, and region. A codebook of the household variables is listed in Table 13.3.

TABLE 13.3 Codebook for household variables

Variable	Description	Value	Label
V2015	Tenure	1	Owned or being bought
		2	Rented for cash
		3	No cash rent
SC214A	Household Income	01	Less than $5,000
		02	$5,000–7,499
		03	$7,500–9,999
		04	$10,000–12,499
		05	$12,500–14,999
		06	$15,000–17,499
		07	$17,500–19,999
		08	$20,000–24,999
		09	$25,000–29,999
		10	$30,000–34,999
		11	$35,000–39,999
		12	$40,000–49,999
		13	$50,000–74,999
		15	$75,000–99,999
		16	$100,000–149,999
		17	$150,000–199,999
		18	$200,000 or more
V2126B	Place Size (Population) Code	00	Not in a place
		13	Population under 10,000
		16	10,000–49,999
		17	50,000–99,999
		18	100,000–249,999
		19	250,000–499,999
		20	500,000–999,999
		21	1,000,000–2,499,999
		22	2,500,000–4,999,999
		23	5,000,000 or more
V2127B	Region	1	Northeast
		2	Midwest
		3	South
		4	West
V2143	Urbanicity	1	Urban
		2	Suburban
		3	Rural

```
hh_vsum_der <- hh_vsum %>%
  mutate(
    Tenure = factor(
      case_when(
        V2015 == 1 ~ "Owned",
        !is.na(V2015) ~ "Rented"
      ),
      levels = c("Owned", "Rented")
    ),
    Urbanicity = factor(
      case_when(
        V2143 == 1 ~ "Urban",
        V2143 == 2 ~ "Suburban",
        V2143 == 3 ~ "Rural"
      ),
      levels = c("Urban", "Suburban", "Rural")
    ),
    SC214A_num = as.numeric(as.character(SC214A)),
    Income = case_when(
      SC214A_num <= 8 ~ "Less than $25,000",
      SC214A_num <= 12 ~ "$25,000--49,999",
      SC214A_num <= 15 ~ "$50,000--99,999",
      SC214A_num <= 17 ~ "$100,000--199,999",
      SC214A_num <= 18 ~ "$200,000 or more"
    ),
    Income = fct_reorder(Income, SC214A_num, .na_rm = FALSE),
    PlaceSize = case_match(
      as.numeric(as.character(V2126B)),
      0 ~ "Not in a place",
      13 ~ "Population under 10,000",
      16 ~ "10,000--49,999",
      17 ~ "50,000--99,999",
      18 ~ "100,000--249,999",
      19 ~ "250,000--499,999",
      20 ~ "500,000--999,999",
      c(21, 22, 23) ~ "1,000,000 or more"
    ),
    PlaceSize = fct_reorder(PlaceSize, as.numeric(V2126B)),
    Region = case_match(
      as.numeric(V2127B),
      1 ~ "Northeast",
      2 ~ "Midwest",
      3 ~ "South",
      4 ~ "West"
```

```
  ),
  Region = fct_reorder(Region, as.numeric(V2127B))
)
```

As before, we want to check to make sure the recoded variables we create match the existing data as expected.

```
hh_vsum_der %>% count(Tenure, V2015)
```

```
## # A tibble: 4 x 3
##    Tenure V2015       n
##    <fct>  <fct>   <int>
## 1 Owned  1      101944
## 2 Rented 2       46269
## 3 Rented 3        1925
## 4 <NA>   <NA>   106322
```

```
hh_vsum_der %>% count(Urbanicity, V2143)
```

```
## # A tibble: 3 x 3
##    Urbanicity V2143       n
##    <fct>      <fct>   <int>
## 1 Urban      1       26878
## 2 Suburban   2      173491
## 3 Rural      3       56091
```

```
hh_vsum_der %>% count(Income, SC214A)
```

```
## # A tibble: 18 x 3
##    Income            SC214A     n
##    <fct>             <fct>  <int>
## 1 Less than $25,000  1      7841
## 2 Less than $25,000  2      2626
## 3 Less than $25,000  3      3949
## 4 Less than $25,000  4      5546
## 5 Less than $25,000  5      5445
## 6 Less than $25,000  6      4821
## 7 Less than $25,000  7      5038
## 8 Less than $25,000  8     11887
## 9 $25,000--49,999    9     11550
## 10 $25,000--49,999   10    13689
## 11 $25,000--49,999   11    13655
## 12 $25,000--49,999   12    23282
```

```
## 13 $50,000--99,999    13     44601
## 14 $50,000--99,999    15     33353
## 15 $100,000--199,999  16     34287
## 16 $100,000--199,999  17     15317
## 17 $200,000 or more   18     16892
## 18 <NA>               <NA>    2681
```

```
hh_vsum_der %>% count(PlaceSize, V2126B)
```

```
## # A tibble: 10 x 3
##    PlaceSize                V2126B      n
##    <fct>                    <fct>   <int>
##  1 Not in a place           0       69484
##  2 Population under 10,000  13       39873
##  3 10,000--49,999           16       53002
##  4 50,000--99,999           17       27205
##  5 100,000--249,999         18       24461
##  6 250,000--499,999         19       13111
##  7 500,000--999,999         20       15194
##  8 1,000,000 or more        21        6167
##  9 1,000,000 or more        22        3857
## 10 1,000,000 or more        23        4106
```

```
hh_vsum_der %>% count(Region, V2127B)
```

```
## # A tibble: 4 x 3
##   Region    V2127B      n
##   <fct>     <fct>   <int>
## 1 Northeast 1       41585
## 2 Midwest   2       74666
## 3 South     3       87783
## 4 West      4       52426
```

13.4.2.2 Person variables

For the person file, we create categories for sex, race/Hispanic origin, age categories, and marital status. A codebook of the household variables is located in Table 13.4. We also merge the household demographics to the person file as well as the design variables (V2117 and V2118).

TABLE 13.4 Codebook for person variables

Variable	Description	Value	Label
V3014	Age		12–90
V3015	Current Marital Status	1	Married
		2	Widowed
		3	Divorced
		4	Separated
		5	Never married
V3018	Sex	1	Male
		2	Female
V3023A	Race	01	White only
		02	Black only
		03	American Indian, Alaska native only
		04	Asian only
		05	Hawaiian/Pacific Islander only
		06	White-Black
		07	White-American Indian
		08	White-Asian
		09	White-Hawaiian
		10	Black-American Indian
		11	Black-Asian
		12	Black-Hawaiian/Pacific Islander
		13	American Indian-Asian
		14	Asian-Hawaiian/Pacific Islander
		15	White-Black-American Indian
		16	White-Black-Asian
		17	White-American Indian-Asian
		18	White-Asian-Hawaiian
		19	2 or 3 races
		20	4 or 5 races
V3024	Hispanic Origin	1	Yes
		2	No

```
NHOPI <- "Native Hawaiian or Other Pacific Islander"

pers_vsum_der <- pers_vsum %>%
  mutate(
    Sex = factor(case_when(
      V3018 == 1 ~ "Male",
      V3018 == 2 ~ "Female"
```

```
  )),
  RaceHispOrigin = factor(
    case_when(
      V3024 == 1 ~ "Hispanic",
      V3023A == 1 ~ "White",
      V3023A == 2 ~ "Black",
      V3023A == 4 ~ "Asian",
      V3023A == 5 ~ NHOPI,
      TRUE ~ "Other"
    ),
    levels = c(
      "White", "Black", "Hispanic",
      "Asian", NHOPI, "Other"
    )
  ),
  V3014_num = as.numeric(as.character(V3014)),
  AgeGroup = case_when(
    V3014_num <= 17 ~ "12--17",
    V3014_num <= 24 ~ "18--24",
    V3014_num <= 34 ~ "25--34",
    V3014_num <= 49 ~ "35--49",
    V3014_num <= 64 ~ "50--64",
    V3014_num <= 90 ~ "65 or older"
  ),
  AgeGroup = fct_reorder(AgeGroup, V3014_num),
  MaritalStatus = factor(
    case_when(
      V3015 == 1 ~ "Married",
      V3015 == 2 ~ "Widowed",
      V3015 == 3 ~ "Divorced",
      V3015 == 4 ~ "Separated",
      V3015 == 5 ~ "Never married"
    ),
    levels = c(
      "Never married", "Married",
      "Widowed", "Divorced",
      "Separated"
    )
  )
) %>%
left_join(
  hh_vsum_der %>% select(
    YEARQ, IDHH,
    V2117, V2118, Tenure:Region
```

```
  ),
  by = c("YEARQ", "IDHH")
)
```

As before, we want to check to make sure the recoded variables we create
match the existing data as expected.

```
pers_vsum_der %>% count(Sex, V3018)
```

```
## # A tibble: 2 x 3
##   Sex      V3018      n
##   <fct>    <fct>  <int>
## 1 Female   2     150956
## 2 Male     1     140922
```

```
pers_vsum_der %>% count(RaceHispOrigin, V3024)
```

```
## # A tibble: 11 x 3
##    RaceHispOrigin                              V3024      n
##    <fct>                                       <fct>  <int>
##  1 White                                       2     197292
##  2 White                                       8        883
##  3 Black                                       2      29947
##  4 Black                                       8        120
##  5 Hispanic                                    1      41450
##  6 Asian                                       2      16015
##  7 Asian                                       8         61
##  8 Native Hawaiian or Other Pacific Islander   2        891
##  9 Native Hawaiian or Other Pacific Islander   8          9
## 10 Other                                       2       5161
## 11 Other                                       8         49
```

```
pers_vsum_der %>%
  filter(RaceHispOrigin != "Hispanic" |
    is.na(RaceHispOrigin)) %>%
  count(RaceHispOrigin, V3023A)
```

```
## # A tibble: 20 x 3
##    RaceHispOrigin    V3023A      n
##    <fct>             <fct>  <int>
##  1 White             1     198175
##  2 Black             2      30067
##  3 Asian             4      16076
```

```
##   4 Native Hawaiian or Other Pacific Islander 5         900
##   5 Other                                    3        1319
##   6 Other                                    6        1217
##   7 Other                                    7        1025
##   8 Other                                    8         837
##   9 Other                                    9         184
## 10 Other                                   10         178
## 11 Other                                   11          87
## 12 Other                                   12          27
## 13 Other                                   13          13
## 14 Other                                   14          53
## 15 Other                                   15         136
## 16 Other                                   16          45
## 17 Other                                   17          11
## 18 Other                                   18          33
## 19 Other                                   19          22
## 20 Other                                   20          23
```

```
pers_vsum_der %>%
  group_by(AgeGroup) %>%
  summarize(
    minAge = min(V3014),
    maxAge = max(V3014),
    .groups = "drop"
  )
```

```
## # A tibble: 6 x 3
##    AgeGroup    minAge maxAge
##    <fct>        <dbl>  <dbl>
## 1 12--17          12     17
## 2 18--24          18     24
## 3 25--34          25     34
## 4 35--49          35     49
## 5 50--64          50     64
## 6 65 or older     65     90
```

```
pers_vsum_der %>% count(MaritalStatus, V3015)
```

```
## # A tibble: 6 x 3
##    MaritalStatus V3015      n
##    <fct>         <fct>  <int>
## 1 Never married 5      90425
## 2 Married       1     148131
## 3 Widowed       2      17668
```

```
## 4 Divorced       3       28596
## 5 Separated      4        4524
## 6 <NA>           8        2534
```

We then create tibbles that contain only the variables we need, which makes it easier to use them for analyses.

```
hh_vsum_slim <- hh_vsum_der %>%
  select(
    YEARQ:V2118,
    WGTVICCY:ADJINC_WT,
    Tenure,
    Urbanicity,
    Income,
    PlaceSize,
    Region
  )

pers_vsum_slim <- pers_vsum_der %>%
  select(YEARQ:WGTPERCY, WGTVICCY:ADJINC_WT, Sex:Region)
```

To calculate estimates about types of crime, such as what percentage of violent crimes are reported to the police, we must use the incident file. The incident file is not guaranteed to have every pseudo-stratum and half-sample code, so dummy records are created to append before estimation. Finally, we merge demographic variables onto the incident tibble.

```
dummy_records <- hh_vsum_slim %>%
  distinct(V2117, V2118) %>%
  mutate(
    Dummy = 1,
    WGTVICCY = 1,
    NEWWGT = 1
  )

inc_analysis <- inc_ind %>%
  mutate(Dummy = 0) %>%
  left_join(select(pers_vsum_slim, YEARQ, IDHH, IDPER, Sex:Region),
    by = c("YEARQ", "IDHH", "IDPER")
  ) %>%
  bind_rows(dummy_records) %>%
  select(
    YEARQ:IDPER,
    WGTVICCY,
```

```
    NEWWGT,
    V4529,
    WeapCat,
    ReportPolice,
    Property:Region
  )
```

The tibbles `hh_vsum_slim`, `pers_vsum_slim`, and `inc_analysis` can now be used
to create design objects and calculate crime rate estimates.

13.5   Survey design objects

All the data preparation above is necessary to create the design objects and
finally begin analysis. We create three design objects for different types of
analysis, depending on the estimate we are creating. For the incident data, the
weight of analysis is `NEWWGT`, which we constructed previously. The household
and person-level data use `WGTHHCY` and `WGTPERCY`, respectively. For all analyses,
`V2117` is the strata variable, and `V2118` is the cluster/PSU variable for analysis.
This information can be found in the User's Guide (Shook-Sa et al., 2015).

```
inc_des <- inc_analysis %>%
  as_survey_design(
    weight = NEWWGT,
    strata = V2117,
    ids = V2118,
    nest = TRUE
  )

hh_des <- hh_vsum_slim %>%
  as_survey_design(
    weight = WGTHHCY,
    strata = V2117,
    ids = V2118,
    nest = TRUE
  )

pers_des <- pers_vsum_slim %>%
  as_survey_design(
    weight = WGTPERCY,
    strata = V2117,
```

```
    ids = V2118,
    nest = TRUE
)
```

13.6 Calculating estimates

Now that we have prepared our data and created the design objects, we can calculate our estimates. As a reminder, those are:

1. Victimization totals estimate the number of criminal victimizations with a given characteristic.

2. Victimization proportions estimate characteristics among victimizations or victims.

3. Victimization rates are estimates of the number of victimizations per 1,000 persons or households in the population.

4. Prevalence rates are estimates of the percentage of the population (persons or households) who are victims of a crime.

13.6.1 Estimation 1: Victimization totals

There are two ways to calculate victimization totals. Using the incident design object (inc_des) is the most straightforward method, but the person (pers_des) and household (hh_des) design objects can be used as well if the adjustment factor (ADJINC_WT) is incorporated. In the example below, the total number of property and violent victimizations is first calculated using the incident file and then using the household and person design objects. The incident file is smaller, and thus, estimation is faster using that file, but the estimates are the same as illustrated in Table 13.5, Table 13.6, and Table 13.7.

```
vt1 <-
  inc_des %>%
  summarize(
    Property_Vzn = survey_total(Property, na.rm = TRUE),
    Violent_Vzn = survey_total(Violent, na.rm = TRUE)
  ) %>%
  gt() %>%
  tab_spanner(
    label = "Property Crime",
```

```
      columns = starts_with("Property")
    ) %>%
    tab_spanner(
      label = "Violent Crime",
      columns = starts_with("Violent")
    ) %>%
    cols_label(
      ends_with("Vzn") ~ "Total",
      ends_with("se") ~ "S.E."
    ) %>%
    fmt_number(decimals = 0)


vt2a <- hh_des %>%
  summarize(Property_Vzn = survey_total(Property * ADJINC_WT,
    na.rm = TRUE
  )) %>%
  gt() %>%
  tab_spanner(
    label = "Property Crime",
    columns = starts_with("Property")
  ) %>%
  cols_label(
    ends_with("Vzn") ~ "Total",
    ends_with("se") ~ "S.E."
  ) %>%
  fmt_number(decimals = 0)

vt2b <- pers_des %>%
  summarize(Violent_Vzn = survey_total(Violent * ADJINC_WT,
    na.rm = TRUE
  )) %>%
  gt() %>%
  tab_spanner(
    label = "Violent Crime",
    columns = starts_with("Violent")
  ) %>%
  cols_label(
    ends_with("Vzn") ~ "Total",
    ends_with("se") ~ "S.E."
  ) %>%
  fmt_number(decimals = 0)
```

TABLE 13.5 Estimates of total property and violent victimizations with standard errors calculated using the incident design object, 2021 (vt1)

Property Crime		Violent Crime	
Total	S.E.	Total	S.E.
11,682,056	263,844	4,598,306	198,115

TABLE 13.6 Estimates of total property victimizations with standard errors calculated using the household design object, 2021 (vt2a)

Property Crime	
Total	S.E.
11,682,056	263,844

TABLE 13.7 Estimates of total violent victimizations with standard errors calculated using the person design object, 2021 (vt2b)

Violent Crime	
Total	S.E.
4,598,306	198,115

The number of victimizations estimated using the incident file is equivalent to the person and household file method. There were an estimated 11,682,056 property victimizations and 4,598,306 violent victimizations in 2021.

13.6.2 Estimation 2: Victimization proportions

Victimization proportions are proportions describing features of a victimization. The key here is that these are estimates among victimizations, not among the population. These types of estimates can only be calculated using the incident design object (inc_des).

For example, we could be interested in the percentage of property victimizations reported to the police as shown in the following code with an estimate, the standard error, and 95% confidence interval:

```
prop1 <- inc_des %>%
  filter(Property) %>%
  summarize(Pct = survey_mean(ReportPolice,
```

```
    na.rm = TRUE,
    proportion = TRUE,
    vartype = c("se", "ci")
  ) * 100)

prop1
```

```
## # A tibble: 1 x 4
##      Pct Pct_se Pct_low Pct_upp
##    <dbl>  <dbl>   <dbl>   <dbl>
## 1  30.8  0.798    29.2    32.4
```

Or, the percentage of violent victimizations that are in urban areas:

```
prop2 <- inc_des %>%
  filter(Violent) %>%
  summarize(Pct = survey_mean(Urbanicity == "Urban",
    na.rm = TRUE
  ) * 100)

prop2
```

```
## # A tibble: 1 x 2
##      Pct Pct_se
##    <dbl>  <dbl>
## 1  18.1   1.49
```

In 2021, we estimate that 30.8% of property crimes were reported to the police, and 18.1% of violent crimes occurred in urban areas.

13.6.3   Estimation 3: Victimization rates

Victimization rates measure the number of victimizations per population. They are not an estimate of the proportion of households or persons who are victimized, which is the prevalence rate described in Section 13.6.4. Victimization rates are estimated using the household (hh_des) or person (pers_des) design objects depending on the type of crime, and the adjustment factor (ADJINC_WT) must be incorporated. We return to the example of property and violent victimizations used in the example for victimization totals (Section 13.6.1). In the following example, the property victimization totals are calculated as above, as well as the property victimization rate (using survey_mean()) and the population size using survey_total().

Victimization rates use the incident weight in the numerator and the person or household weight in the denominator. This is accomplished by calculating the rates with the weight adjustment (ADJINC_WT) multiplied by the estimate of interest. Let's look at an example of property victimization.

```
vr_prop <- hh_des %>%
  summarize(
    Property_Vzn = survey_total(Property * ADJINC_WT,
      na.rm = TRUE
    ),
    Property_Rate = survey_mean(Property * ADJINC_WT * 1000,
      na.rm = TRUE
    ),
    PopSize = survey_total(1, vartype = NULL)
  )
vr_prop
```

```
## # A tibble: 1 x 5
##   Property_Vzn Property_Vzn_se Property_Rate Property_Rate_se  PopSize
##          <dbl>           <dbl>         <dbl>            <dbl>    <dbl>
## 1    11682056.         263844.          90.3             1.95   1.29e8
```

In the output above, we see the estimate for property victimization rate in 2021 was 90.3 per 1,000 households. This is consistent with calculating the number of victimizations per 1,000 population, as demonstrated in the following code output.

```
vr_prop %>%
  select(-ends_with("se")) %>%
  mutate(Property_Rate_manual = Property_Vzn / PopSize * 1000)
```

```
## # A tibble: 1 x 4
##   Property_Vzn Property_Rate    PopSize Property_Rate_manual
##          <dbl>         <dbl>      <dbl>                <dbl>
## 1    11682056.          90.3 129319232.                 90.3
```

Victimization rates can also be calculated based on particular characteristics of the victimization. In the following example, we calculate the rate of aggravated assault with no weapon, firearm, knife, and another weapon.

```
pers_des %>%
  summarize(across(
    starts_with("AAST_"),
    ~ survey_mean(. * ADJINC_WT * 1000, na.rm = TRUE)
  ))
```

```
## # A tibble: 1 x 8
##    AAST_NoWeap AAST_NoWeap_se AAST_Firearm AAST_Firearm_se AAST_Knife
##          <dbl>          <dbl>        <dbl>           <dbl>      <dbl>
## 1        0.249         0.0595        0.860           0.101      0.455
## # i 3 more variables: AAST_Knife_se <dbl>, AAST_Other <dbl>,
## #    AAST_Other_se <dbl>
```

A common desire is to calculate victimization rates by several characteristics. For example, we may want to calculate the violent victimization rate and aggravated assault rate by sex, race/Hispanic origin, age group, marital status, and household income. This requires a separate group_by() statement for each categorization. Thus, we make a function to do this and then use the map_df() function from the {purrr} package to loop through the variables (Wickham and Henry, 2023). This function takes a demographic variable as its input (byarvar) and calculates the violent and aggravated assault victimization rate for each level. It then creates some columns with the variable, the level of each variable, and a numeric version of the variable (LevelNum) for sorting later. The function is run across multiple variables using map() and then stacks the results into a single output using bind_rows().

```
pers_est_by <- function(byvar) {
  pers_des %>%
    rename(Level := {{ byvar }}) %>%
    filter(!is.na(Level)) %>%
    group_by(Level) %>%
    summarize(
      Violent = survey_mean(Violent * ADJINC_WT * 1000, na.rm = TRUE),
      AAST = survey_mean(AAST * ADJINC_WT * 1000, na.rm = TRUE)
    ) %>%
    mutate(
      Variable = byvar,
      LevelNum = as.numeric(Level),
      Level = as.character(Level)
    ) %>%
    select(Variable, Level, LevelNum, everything())
}

pers_est_df <-
  c("Sex", "RaceHispOrigin", "AgeGroup", "MaritalStatus", "Income") %>%
  map(pers_est_by) %>%
  bind_rows()
```

The output from all the estimates is cleaned to create better labels, such as going from "RaceHispOrigin" to "Race/Hispanic Origin." Finally, the {gt} package is used to make a publishable table (Table 13.8). Using the functions from the {gt} package, we add column labels and footnotes and present estimates rounded to the first decimal place (Iannone et al., 2024).

```
vr_gt <- pers_est_df %>%
  mutate(
    Variable = case_when(
      Variable == "RaceHispOrigin" ~ "Race/Hispanic Origin",
      Variable == "MaritalStatus" ~ "Marital Status",
      Variable == "AgeGroup" ~ "Age",
      TRUE ~ Variable
    )
  ) %>%
  select(-LevelNum) %>%
  group_by(Variable) %>%
  gt(rowname_col = "Level") %>%
  tab_spanner(
    label = "Violent Crime",
    id = "viol_span",
    columns = c("Violent", "Violent_se")
  ) %>%
  tab_spanner(
    label = "Aggravated Assault",
    columns = c("AAST", "AAST_se")
  ) %>%
  cols_label(
    Violent = "Rate",
    Violent_se = "S.E.",
    AAST = "Rate",
    AAST_se = "S.E.",
  ) %>%
  fmt_number(
    columns = c("Violent", "Violent_se", "AAST", "AAST_se"),
    decimals = 1
  ) %>%
  tab_footnote(
    footnote = "Includes rape or sexual assault, robbery,
    aggravated assault, and simple assault.",
    locations = cells_column_spanners(spanners = "viol_span")
  ) %>%
  tab_footnote(
    footnote = "Excludes persons of Hispanic origin.",
    locations =
      cells_stub(rows = Level %in%
        c("White", "Black", "Asian", NHOPI, "Other"))
  ) %>%
  tab_footnote(
    footnote = "Includes persons who identified as
    Native Hawaiian or Other Pacific Islander only.",
```

```
    locations = cells_stub(rows = Level == NHOPI)
) %>%
tab_footnote(
    footnote = "Includes persons who identified as American Indian or
    Alaska Native only or as two or more races.",
    locations = cells_stub(rows = Level == "Other")
) %>%
tab_source_note(
    source_note = md("*Note*: Rates per 1,000 persons age 12 or older.")
) %>%
tab_source_note(
    source_note = md("*Source*: Bureau of Justice Statistics,
                National Crime Victimization Survey, 2021.")
) %>%
tab_stubhead(label = "Victim Demographic") %>%
tab_caption("Rate and standard error of violent victimization,
            by type of crime and demographic characteristics, 2021")
```

vr_gt

TABLE 13.8 Rate and standard error of violent victimization, by type of crime and demographic characteristics, 2021

Victim Demographic	Violent Crime[1]		Aggravated Assault	
	Rate	S.E.	Rate	S.E.
Sex				
Female	15.5	0.9	2.3	0.2
Male	17.5	1.1	3.2	0.3
Race/Hispanic Origin				
White[2]	16.1	0.9	2.7	0.3
Black[2]	18.5	2.2	3.7	0.7
Hispanic	15.9	1.7	2.3	0.4
Asian[2]	8.6	1.3	1.9	0.6
Native Hawaiian or Other Pacific Islander[2,3]	36.1	34.4	0.0	0.0
Other[2,4]	45.4	13.0	6.2	2.0
Age				
12–17	13.2	2.2	2.5	0.8
18–24	23.1	2.1	3.9	0.9
25–34	22.0	2.1	4.0	0.6
35–49	19.4	1.6	3.6	0.5

50–64	16.9	1.9	2.0	0.3
65 or older	6.4	1.1	1.1	0.3
Marital Status				
Never married	22.2	1.4	4.0	0.4
Married	9.5	0.9	1.5	0.2
Widowed	10.7	3.5	0.9	0.2
Divorced	27.4	2.9	4.0	0.7
Separated	36.8	6.7	8.8	3.1
Income				
Less than $25,000	29.6	2.5	5.1	0.7
$25,000–49,999	16.9	1.5	3.0	0.4
$50,000–99,999	14.6	1.1	1.9	0.3
$100,000–199,999	12.2	1.3	2.5	0.4
$200,000 or more	9.7	1.4	1.7	0.6

[1] Includes rape or sexual assault, robbery, aggravated assault, and simple assault.
[2] Excludes persons of Hispanic origin.
[3] Includes persons who identified as Native Hawaiian or Other Pacific Islander only.
[4] Includes persons who identified as American Indian or Alaska Native only or as two or more races.
Note: Rates per 1,000 persons age 12 or older.
Source: Bureau of Justice Statistics, National Crime Victimization Survey, 2021.

13.6.4   Estimation 4: Prevalence rates

Prevalence rates differ from victimization rates, as the numerator is the number of people or households victimized rather than the number of victimizations. To calculate the prevalence rates, we must run another summary of the data by calculating an indicator for whether a person or household is a victim of a particular crime at any point in the year. Below is an example of calculating the indicator and then the prevalence rate of violent crime and aggravated assault.

```
pers_prev_des <-
  pers_vsum_slim %>%
  mutate(Year = floor(YEARQ)) %>%
  mutate(
    Violent_Ind = sum(Violent) > 0,
    AAST_Ind = sum(AAST) > 0,
    .by = c("Year", "IDHH", "IDPER")
```

```
) %>%
as_survey(
  weight = WGTPERCY,
  strata = V2117,
  ids = V2118,
  nest = TRUE
)

pers_prev_ests <- pers_prev_des %>%
  summarize(
    Violent_Prev = survey_mean(Violent_Ind * 100),
    AAST_Prev = survey_mean(AAST_Ind * 100)
  )

pers_prev_ests
```

```
## # A tibble: 1 x 4
##   Violent_Prev Violent_Prev_se AAST_Prev AAST_Prev_se
##          <dbl>           <dbl>     <dbl>        <dbl>
## 1        0.980          0.0349     0.215       0.0143
```

In the example above, the indicator is multiplied by 100 to return a percentage rather than a proportion. In 2021, we estimate that 0.98% of people aged 12 and older were victims of violent crime in the United States, and 0.22% were victims of aggravated assault.

13.7   Statistical testing

For any of the types of estimates discussed, we can also perform statistical testing. For example, we could test whether property victimization rates are different between properties that are owned versus rented. First, we calculate the point estimates.

```
prop_tenure <- hh_des %>%
  group_by(Tenure) %>%
  summarize(
    Property_Rate = survey_mean(Property * ADJINC_WT * 1000,
      na.rm = TRUE, vartype = "ci"
    ),
  )

prop_tenure
```

```
## # A tibble: 3 x 4
##    Tenure Property_Rate Property_Rate_low Property_Rate_upp
##    <fct>         <dbl>            <dbl>            <dbl>
## 1 Owned          68.2             64.3             72.1
## 2 Rented        130.             123.             137.
## 3 <NA>           NaN              NaN              NaN
```

The property victimization rate for rented households is 129.8 per 1,000 households, while the property victimization rate for owned households is 68.2, which seem very different, especially given the non-overlapping confidence intervals. However, survey data are inherently non-independent, so statistical testing cannot be done by comparing confidence intervals. To conduct the statistical test, we first need to create a variable that incorporates the adjusted incident weight (ADJINC_WT), and then the test can be conducted on this adjusted variable as discussed in Chapter 6.

```
prop_tenure_test <- hh_des %>%
  mutate(
    Prop_Adj = Property * ADJINC_WT * 1000
  ) %>%
  svyttest(
    formula = Prop_Adj ~ Tenure,
    design = .,
    na.rm = TRUE
  ) %>%
  broom::tidy()
```

```
prop_tenure_test %>%
  mutate(p.value = pretty_p_value(p.value)) %>%
  gt() %>%
  fmt_number()
```

TABLE 13.9 T-test output for estimates of property victimization rates between properties that are owned versus rented, NCVS 2021

estimate	statistic	p.value	parameter	conf.low	conf.high	method	alternative
61.62	16.04	<0.0001	169.00	54.03	69.21	Design-based t-test	two.sided

The output of the statistical test shown in Table 13.9 indicates a difference of 61.6 between the property victimization rates of renters and owners, and the test is highly significant with the p-value of <0.0001.

13.8   Exercises

1. What proportion of completed motor vehicle thefts are not reported to the police? Hint: Use the codebook to look at the definition of Type of Crime (V4529).

2. How many violent crimes occur in each region?

3. What is the property victimization rate among each income level?

4. What is the difference between the violent victimization rate between males and females? Is it statistically different?

14

AmericasBarometer vignette

> ### Prerequisites
>
> For this chapter, load the following packages:
>
> ```
> library(tidyverse)
> library(survey)
> library(srvyr)
> library(sf)
> library(rnaturalearth)
> library(rnaturalearthdata)
> library(gt)
> library(ggpattern)
> ```
>
> This vignette uses a subset of data from the 2021 AmericasBarometer survey. Download the raw files, available on the LAPOP website[a]. We work with version 1.2 of the data, and there are separate files for each of the 22 countries. To import all files into R while ignoring the Stata labels, we recommend running the following code using the read_stata() function from the {haven} package (Wickham et al., 2023b):


```
stata_files <- list.files(here("RawData", "LAPOP_2021"), "*.dta")

read_stata_unlabeled <- function(file) {
  read_stata(file) %>%
    zap_labels() %>%
    zap_label()
}

ambarom_in <- here("RawData", "LAPOP_2021", stata_files) %>%
  map_df(read_stata_unlabeled) %>%
  select(pais, strata, upm, weight1500, strata, core_a_core_b,
         q2, q1tb, covid2at, a4, idio2, idio2cov, it1, jc13,
         m1, mil10a, mil10e, ccch1, ccch3, ccus1, ccus3,
         edr, ocup4a, q14, q11n, q12c, q12bn,
         starts_with("covidedu1"), gi0n,
         r15, r18n, r18)
```

The code above reads all the .dta files and combines them into one
tibble.

---

^ahttp://datasets.americasbarometer.org/database/index.php

14.1   Introduction

The AmericasBarometer surveys, conducted by the LAPOP Lab (LAPOP,
2023b), are public opinion surveys of the Americas focused on democracy. The
study was launched in 2004/2005 with 11 countries. Though the participating
countries change over time, AmericasBarometer maintains a consistent methodology across many of them. In 2021, the study included 22 countries ranging
from Canada in the north to Chile and Argentina in the south (LAPOP,
2023a).

Historically, surveys were administered through in-person household interviews,
but the COVID-19 pandemic changed the study significantly. Now, random-
digit dialing (RDD) of mobile phones is used in all countries except the United
States and Canada (LAPOP, 2021c). In Canada, LAPOP collaborated with
the Environics Institute to collect data from a panel of Canadians using a
web survey (LAPOP, 2021a). In the United States, YouGov conducted a web
survey on behalf of LAPOP among its panelists (LAPOP, 2021b).

The survey includes a core set of questions for all countries, but not every
question is asked in each country. Additionally, some questions are only posed

to half of the respondents in a country, with different randomized sections (LAPOP, 2021d).

14.2 Data structure

Each country and year has its own file available in Stata format (.dta). In this vignette, we download and combine all the data from the 22 participating countries in 2021. We subset the data to a smaller set of columns, as noted in the Prerequisites box. We recommend reviewing the core questionnaire to understand the common variables across the countries (LAPOP, 2021d).

14.3 Preparing files

Many of the variables are coded as numeric and do not have intuitive variable names, so the next step is to create derived variables and wrangle the data for analysis. Using the core questionnaire as a codebook, we reference the factor descriptions to create derived variables with informative names:

```
ambarom <- ambarom_in %>%
  mutate(
    Country = factor(
      case_match(
        pais,
        1 ~ "Mexico",
        2 ~ "Guatemala",
        3 ~ "El Salvador",
        4 ~ "Honduras",
        5 ~ "Nicaragua",
        6 ~ "Costa Rica",
        7 ~ "Panama",
        8 ~ "Colombia",
        9 ~ "Ecuador",
        10 ~ "Bolivia",
        11 ~ "Peru",
        12 ~ "Paraguay",
        13 ~ "Chile",
        14 ~ "Uruguay",
        15 ~ "Brazil",
```

```
         17 ~ "Argentina",
         21 ~ "Dominican Republic",
         22 ~ "Haiti",
         23 ~ "Jamaica",
         24 ~ "Guyana",
         40 ~ "United States",
         41 ~ "Canada"
       )
     ),
     CovidWorry = fct_reorder(
       case_match(
         covid2at,
         1 ~ "Very worried",
         2 ~ "Somewhat worried",
         3 ~ "A little worried",
         4 ~ "Not worried at all"
       ),
       covid2at,
       .na_rm = FALSE
     )
   ) %>%
   rename(
     Educ_NotInSchool = covidedu1_1,
     Educ_NormalSchool = covidedu1_2,
     Educ_VirtualSchool = covidedu1_3,
     Educ_Hybrid = covidedu1_4,
     Educ_NoSchool = covidedu1_5,
     BroadbandInternet = r18n,
     Internet = r18
   )
```

At this point, it is a good time to check the cross-tabs between the original and newly derived variables. These tables help us confirm that we have correctly matched the numeric data from the original dataset to the renamed factor data in the new dataset. For instance, let's check the original variable pais and the derived variable Country. We can consult the questionnaire or codebook to confirm that Argentina is coded as 17, Bolivia as 10, etc. Similarly, for CovidWorry and covid2at, we can verify that Very worried is coded as 1, and so on for the other variables.

```
ambarom %>%
  count(Country, pais) %>%
  print(n = 22)
```

```
## # A tibble: 22 x 3
##    Country               pais     n
##    <fct>                <dbl> <int>
##  1 Argentina               17  3011
##  2 Bolivia                 10  3002
##  3 Brazil                  15  3016
##  4 Canada                  41  2201
##  5 Chile                   13  2954
##  6 Colombia                 8  2993
##  7 Costa Rica               6  2977
##  8 Dominican Republic      21  3000
##  9 Ecuador                  9  3005
## 10 El Salvador              3  3245
## 11 Guatemala                2  3000
## 12 Guyana                  24  3011
## 13 Haiti                   22  3088
## 14 Honduras                 4  2999
## 15 Jamaica                 23  3121
## 16 Mexico                   1  2998
## 17 Nicaragua                5  2997
## 18 Panama                   7  3183
## 19 Paraguay                12  3004
## 20 Peru                    11  3038
## 21 United States           40  1500
## 22 Uruguay                 14  3009
```

```
ambarom %>%
  count(CovidWorry, covid2at)
```

```
## # A tibble: 5 x 3
##   CovidWorry         covid2at     n
##   <fct>                 <dbl> <int>
## 1 Very worried              1 24327
## 2 Somewhat worried          2 13233
## 3 A little worried          3 11478
## 4 Not worried at all        4  8628
## 5 <NA>                     NA  6686
```

14.4 Survey design objects

The technical report is the best reference for understanding how to specify
the sampling design in R (LAPOP, 2021c). The data include two weights:

wt and `weight1500`. The first weight variable is specific to each country and sums to the sample size, but it is calibrated to reflect each country's demographics. The second weight variable sums to 1500 for each country and is recommended for multi-country analyses. Although not explicitly stated in the documentation, the Stata syntax example (`svyset upm [pw=weight1500], strata(strata)`) indicates the variable `upm` is a clustering variable, and `strata` is the strata variable. Therefore, the design object for multi-country analysis is created in R as follows:

```
ambarom_des <- ambarom %>%
  as_survey_design(
    ids = upm,
    strata = strata,
    weight = weight1500
  )
```

One interesting thing to note is that these weight variables can provide estimates for comparing countries but not for multi-country estimates. This is due to the fact that the weights do not account for the different sizes of countries. For example, Canada has about 10% of the population of the United States, but an estimate that uses records from both countries would weigh them equally.

14.5   Calculating estimates

When calculating estimates from the data, we use the survey design object `ambarom_des` and then apply the `survey_mean()` function. The next sections walk through a few examples.

14.5.1   Example: Worry about COVID-19

This survey was administered between March and August 2021, with the specific timing varying by country[1]. Given the state of the pandemic at that time, several questions about COVID-19 were included. According to the core questionnaire (LAPOP, 2021d), the first question asked about COVID-19 was:

---

[1]See table 2 in LAPOP (2021c) for dates by country

> How worried are you about the possibility that you or someone in your household will get sick from coronavirus in the next 3 months?
>
> - Very worried
> - Somewhat worried
> - A little worried
> - Not worried at all

If we are interested in those who are very worried or somewhat worried, we can create a new variable (CovidWorry_bin) that groups levels of the original question using the fct_collapse() function from the {forcats} package (Wickham, 2023). We then use the survey_count() function to understand how responses are distributed across each category of the original variable (CovidWorry) and the new variable (CovidWorry_bin).

```
covid_worry_collapse <- ambarom_des %>%
  mutate(CovidWorry_bin = fct_collapse(
    CovidWorry,
    WorriedHi = c("Very worried", "Somewhat worried"),
    WorriedLo = c("A little worried", "Not worried at all")
  ))

covid_worry_collapse %>%
  survey_count(CovidWorry_bin, CovidWorry)
```

```
## # A tibble: 5 x 4
##   CovidWorry_bin CovidWorry             n  n_se
##   <fct>          <fct>              <dbl> <dbl>
## 1 WorriedHi      Very worried       12369.  83.6
## 2 WorriedHi      Somewhat worried    6378.  63.4
## 3 WorriedLo      A little worried    5896.  62.6
## 4 WorriedLo      Not worried at all  4840.  59.7
## 5 <NA>           <NA>                3518.  42.2
```

With this new variable, we can now use survey_mean() to calculate the percentage of people in each country who are either very or somewhat worried about COVID-19. There are missing data, as indicated in the survey_count() output above, so we need to use na.rm = TRUE in the survey_mean() function to handle the missing values.

```
covid_worry_country_ests <- covid_worry_collapse %>%
  group_by(Country) %>%
  summarize(p = survey_mean(CovidWorry_bin == "WorriedHi",
    na.rm = TRUE
  ) * 100)

covid_worry_country_ests
```

```
## # A tibble: 22 x 3
##    Country                 p  p_se
##    <fct>               <dbl> <dbl>
##  1 Argentina            65.8  1.08
##  2 Bolivia              71.6  0.960
##  3 Brazil               83.5  0.962
##  4 Canada               48.9  1.34
##  5 Chile                81.8  0.828
##  6 Colombia             67.9  1.12
##  7 Costa Rica           72.6  0.952
##  8 Dominican Republic   50.1  1.13
##  9 Ecuador              71.7  0.967
## 10 El Salvador          52.5  1.02
## # i 12 more rows
```

To view the results for all countries, we can use the {gt} package to create Table 14.1 (Iannone et al., 2024).

```
covid_worry_country_ests_gt <- covid_worry_country_ests %>%
  gt(rowname_col = "Country") %>%
  cols_label(
    p = "%",
    p_se = "S.E."
  ) %>%
  fmt_number(decimals = 1) %>%
  tab_source_note(md("*Source*: AmericasBarometer Surveys, 2021"))

covid_worry_country_ests_gt
```

14.5.2 Example: Education affected by COVID-19

In the core questionnaire (LAPOP, 2021d), respondents were also asked a question about how the pandemic affected education. This question was asked to households with children under the age of 13, and respondents could select more than one option, as follows:

TABLE 14.1 Percentage worried about the possibility that they or someone in their household will get sick from coronavirus in the next 3 months

	%	S.E.
Argentina	65.8	1.1
Bolivia	71.6	1.0
Brazil	83.5	1.0
Canada	48.9	1.3
Chile	81.8	0.8
Colombia	67.9	1.1
Costa Rica	72.6	1.0
Dominican Republic	50.1	1.1
Ecuador	71.7	1.0
El Salvador	52.5	1.0
Guatemala	69.3	1.0
Guyana	60.0	1.6
Haiti	54.4	1.8
Honduras	64.6	1.1
Jamaica	28.4	0.9
Mexico	63.6	1.0
Nicaragua	80.0	1.0
Panama	70.2	1.0
Paraguay	61.5	1.1
Peru	77.1	2.5
United States	46.6	1.7
Uruguay	60.9	1.1

Source: AmericasBarometer Surveys, 2021

Did any of these children have their school education affected due to the pandemic?

- No, because they are not yet school age or because they do not attend school for another reason
- No, their classes continued normally
- Yes, they went to virtual or remote classes
- Yes, they switched to a combination of virtual and in-person classes
- Yes, they cut all ties with the school

Working with multiple-choice questions can be both challenging and interesting. Let's walk through how to analyze this question. If we are interested in the impact on education, we should focus on the data of those whose children are attending school. This means we need to exclude those who selected the first response option: "No, because they are not yet school age or because they do not attend school for another reason." To do this, we use the Educ_NotInSchool variable in the dataset, which has values of 0 and 1. A value of 1 indicates that the respondent chose the first response option (none of the children are in school), and a value of 0 means that at least one of their children is in school. By filtering the data to those with a value of 0 (they have at least one child in school), we can consider only respondents with at least one child attending school.

Now, let's review the data for those who selected one of the next three response options:

- No, their classes continued normally: Educ_NormalSchool
- Yes, they went to virtual or remote classes: Educ_VirtualSchool
- Yes, they switched to a combination of virtual and in-person classes: Educ_Hybrid

The unweighted cross-tab for these responses is included below. It reveals a wide range of impacts, where many combinations of effects on education are possible.

```
ambarom %>%
  filter(Educ_NotInSchool == 0) %>%
  count(
    Educ_NormalSchool,
    Educ_VirtualSchool,
    Educ_Hybrid
  )
```

```
## # A tibble: 8 x 4
##   Educ_NormalSchool Educ_VirtualSchool Educ_Hybrid       n
##               <dbl>              <dbl>       <dbl>   <int>
## 1                 0                  0           0     861
## 2                 0                  0           1    1192
## 3                 0                  1           0    7554
## 4                 0                  1           1     280
## 5                 1                  0           0     833
## 6                 1                  0           1      18
## 7                 1                  1           0      72
## 8                 1                  1           1       7
```

In reviewing the survey question, we might be interested in knowing the answers to the following:

- What percentage of households indicated that school continued as normal with no virtual or hybrid option?
- What percentage of households indicated that the education medium was changed to either virtual or hybrid?
- What percentage of households indicated that they cut ties with their school?

To find the answers, we create indicators for the first two questions, make national estimates for all three questions, and then construct a summary table for easy viewing. First, we create and inspect the indicators and their distributions using `survey_count()`.

```
ambarom_des_educ <- ambarom_des %>%
  filter(Educ_NotInSchool == 0) %>%
  mutate(
    Educ_OnlyNormal = (Educ_NormalSchool == 1 &
      Educ_VirtualSchool == 0 &
      Educ_Hybrid == 0),
    Educ_MediumChange = (Educ_VirtualSchool == 1 |
      Educ_Hybrid == 1)
  )

ambarom_des_educ %>%
  survey_count(
    Educ_OnlyNormal,
    Educ_NormalSchool,
    Educ_VirtualSchool,
    Educ_Hybrid
  )
```

```
## # A tibble: 8 x 6
##    Educ_OnlyNormal Educ_NormalSchool Educ_VirtualSchool Educ_Hybrid
##    <lgl>                       <dbl>              <dbl>       <dbl>
## 1 FALSE                           0                  0           0
## 2 FALSE                           0                  0           1
## 3 FALSE                           0                  1           0
## 4 FALSE                           0                  1           1
## 5 FALSE                           1                  0           1
## 6 FALSE                           1                  1           0
## 7 FALSE                           1                  1           1
## 8 TRUE                            1                  0           0
## # i 2 more variables: n <dbl>, n_se <dbl>
```

```
ambarom_des_educ %>%
  survey_count(
    Educ_MediumChange,
    Educ_VirtualSchool,
    Educ_Hybrid
  )
```

```
## # A tibble: 4 x 5
##   Educ_MediumChange Educ_VirtualSchool Educ_Hybrid     n  n_se
##   <lgl>                          <dbl>       <dbl> <dbl> <dbl>
## 1 FALSE                              0           0  880. 26.1
## 2 TRUE                               0           1  561. 19.2
## 3 TRUE                               1           0 3812. 49.4
## 4 TRUE                               1           1  136.  9.86
```

Next, we group the data by country and calculate the population estimates for our three questions.

```
covid_educ_ests <-
  ambarom_des_educ %>%
  group_by(Country) %>%
  summarize(
    p_onlynormal = survey_mean(Educ_OnlyNormal, na.rm = TRUE) * 100,
    p_mediumchange = survey_mean(Educ_MediumChange, na.rm = TRUE) * 100,
    p_noschool = survey_mean(Educ_NoSchool, na.rm = TRUE) * 100,
  )

covid_educ_ests
```

```
## # A tibble: 16 x 7
##    Country            p_onlynormal p_onlynormal_se p_mediumchange
##    <fct>                     <dbl>           <dbl>          <dbl>
##  1 Argentina                 5.39            1.14           87.1
##  2 Brazil                    4.28            1.17           81.5
##  3 Chile                     0.715           0.267          96.2
##  4 Colombia                  2.84            0.727          90.3
##  5 Dominican Republic        3.75            0.793          87.4
##  6 Ecuador                   5.18            0.963          87.5
##  7 El Salvador               2.92            0.680          85.8
##  8 Guatemala                 3.00            0.727          82.2
##  9 Guyana                    3.34            0.702          85.3
## 10 Haiti                    81.1             2.25            7.25
```

```
## 11 Honduras                         3.68              0.882          80.7
## 12 Jamaica                          5.42              0.950          88.1
## 13 Panama                           7.20              1.18           89.4
## 14 Paraguay                         4.66              0.939          90.7
## 15 Peru                             2.04              0.604          91.8
## 16 Uruguay                          8.60              1.40           84.3
## # i 3 more variables: p_mediumchange_se <dbl>, p_noschool <dbl>,
## #   p_noschool_se <dbl>
```

Finally, to view the results for all countries, we can use the {gt} package to construct Table 14.2.

```
covid_educ_ests_gt <- covid_educ_ests %>%
  gt(rowname_col = "Country") %>%
  cols_label(
    p_onlynormal = "%",
    p_onlynormal_se = "S.E.",
    p_mediumchange = "%",
    p_mediumchange_se = "S.E.",
    p_noschool = "%",
    p_noschool_se = "S.E."
  ) %>%
  tab_spanner(
    label = "Normal School Only",
    columns = c("p_onlynormal", "p_onlynormal_se")
  ) %>%
  tab_spanner(
    label = "Medium Change",
    columns = c("p_mediumchange", "p_mediumchange_se")
  ) %>%
  tab_spanner(
    label = "Cut Ties with School",
    columns = c("p_noschool", "p_noschool_se")
  ) %>%
  fmt_number(decimals = 1) %>%
  tab_source_note(md("*Source*: AmericasBarometer Surveys, 2021"))

covid_educ_ests_gt
```

TABLE 14.2 Impact on education in households with children under the age of 13 who generally attend school

	Normal School Only		Medium Change		Cut Ties with School	
	%	S.E.	%	S.E.	%	S.E.
Argentina	5.4	1.1	87.1	1.7	9.9	1.6
Brazil	4.3	1.2	81.5	2.3	22.1	2.5
Chile	0.7	0.3	96.2	1.0	4.0	1.0
Colombia	2.8	0.7	90.3	1.4	7.5	1.3
Dominican Re-public	3.8	0.8	87.4	1.5	10.5	1.4
Ecuador	5.2	1.0	87.5	1.4	7.9	1.1
El Salvador	2.9	0.7	85.8	1.5	11.8	1.4
Guatemala	3.0	0.7	82.2	1.7	17.7	1.8
Guyana	3.3	0.7	85.3	1.7	13.0	1.6
Haiti	81.1	2.3	7.2	1.5	11.7	1.8
Honduras	3.7	0.9	80.7	1.7	16.9	1.6
Jamaica	5.4	0.9	88.1	1.4	7.5	1.2
Panama	7.2	1.2	89.4	1.4	3.8	0.9
Paraguay	4.7	0.9	90.7	1.4	6.4	1.2
Peru	2.0	0.6	91.8	1.2	6.8	1.1
Uruguay	8.6	1.4	84.3	2.0	8.0	1.6

Source: AmericasBarometer Surveys, 2021

In the countries that were asked this question, many households experienced a change in their child's education medium. However, in Haiti, only 7.2% of households with children switched to virtual or hybrid learning.

14.6 Mapping survey data

While the table effectively presents the data, a map could also be insightful. To create a map of the countries, we can use the package {rnaturalearth} and subset North and South America with the ne_countries() function (Massicotte and South, 2023). The function returns a simple features (sf) object with many columns (Pebesma and Bivand, 2023), but most importantly, soverignt (sovereignty), geounit (country or territory), and geometry (the shape). For an example of the difference between sovereignty and country/territory, the United States, Puerto Rico, and the U.S. Virgin Islands are all separate units with the same sovereignty. A map without data is plotted in Figure 14.1 using `geom_sf()` from the {ggplot2} package, which plots sf objects (Wickham, 2016).

```
country_shape <-
  ne_countries(
    scale = "medium",
    returnclass = "sf",
    continent = c("North America", "South America")
  )

country_shape %>%
  ggplot() +
  geom_sf()
```

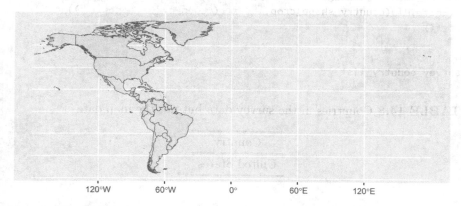

FIGURE 14.1 Map of North and South America

The map in Figure 14.1 appears very wide due to the Aleutian Islands in Alaska extending into the Eastern Hemisphere. We can crop the shapefile to include only the Western Hemisphere using `st_crop()` from the {sf} package, which removes some of the trailing islands of Alaska.

```
country_shape_crop <- country_shape %>%
  st_crop(c(
    xmin = -180,
    xmax = 0,
    ymin = -90,
    ymax = 90
  ))
```

Now that we have the necessary shape files, our next step is to match our survey data to the map. Countries can be named differently (e.g., "U.S.", "U.S.A.", "United States"). To make sure we can visualize our survey data on the map, we need to match the country names in both the survey data and the map data. To do this, we can use the `anti_join()` function from the {dplyr} package to identify the countries in the survey data that are not in the map data. Table 14.3 shows the countries in the survey data but not the map data, and Table 14.4 shows the countries in the map data but not the survey data. As shown below, the United States is referred to as "United States" in the survey data but "United States of America" in the map data.

```
survey_country_list <- ambarom %>% distinct(Country)

survey_country_list_gt <- survey_country_list %>%
  anti_join(country_shape_crop, by = c("Country" = "geounit")) %>%
  gt()
```

```
survey_country_list_gt
```

TABLE 14.3 Countries in the survey data but not the map data

Country
United States

```
map_country_list_gt <- country_shape_crop %>%
  as_tibble() %>%
  select(geounit, sovereignt) %>%
  anti_join(survey_country_list, by = c("geounit" = "Country")) %>%
  arrange(geounit) %>%
  gt()
```

```
map_country_list_gt
```

TABLE 14.4 Countries in the map data but not the survey data

geounit	sovereignt
Anguilla	United Kingdom
Antigua and Barbuda	Antigua and Barbuda
Aruba	Netherlands
Barbados	Barbados
Belize	Belize
Bermuda	United Kingdom
British Virgin Islands	United Kingdom
Cayman Islands	United Kingdom
Cuba	Cuba
Curaçao	Netherlands
Dominica	Dominica
Falkland Islands	United Kingdom
Greenland	Denmark
Grenada	Grenada
Montserrat	United Kingdom
Puerto Rico	United States of America
Saint Barthelemy	France
Saint Kitts and Nevis	Saint Kitts and Nevis
Saint Lucia	Saint Lucia
Saint Martin	France
Saint Pierre and Miquelon	France
Saint Vincent and the Grenadines	Saint Vincent and the Grenadines
Sint Maarten	Netherlands
Suriname	Suriname
The Bahamas	The Bahamas
Trinidad and Tobago	Trinidad and Tobago
Turks and Caicos Islands	United Kingdom
United States Virgin Islands	United States of America
United States of America	United States of America
Venezuela	Venezuela

There are several ways to fix the mismatched names for a successful join. The simplest solution is to rename the data in the shape object before merging. Since only one country name in the survey data differs from the map data, we rename the map data accordingly.

```
country_shape_upd <- country_shape_crop %>%
  mutate(geounit = if_else(geounit == "United States of America",
    "United States", geounit
  ))
```

Now that the country names match, we can merge the survey and map data and then plot the resulting dataset. We begin with the map file and merge it with the survey estimates generated in Section 14.5 (`covid_worry_country_ests` and `covid_educ_ests`). We use the {dplyr} function of `full_join()`, which joins the rows in the map data and the survey estimates based on the columns `geounit` and `Country`. A full join keeps all the rows from both datasets, matching rows when possible. For any rows without matches, the function fills in an `NA` for the missing value (Pebesma and Bivand, 2023).

```
covid_sf <- country_shape_upd %>%
  full_join(covid_worry_country_ests,
    by = c("geounit" = "Country")
  ) %>%
  full_join(covid_educ_ests,
    by = c("geounit" = "Country")
  )
```

After the merge, we create two figures that display the population estimates for the percentage of people worried about COVID-19 (Figure 14.2) and the percentage of households with at least one child participating in virtual or hybrid learning (Figure 14.3). We also add a crosshatch pattern to the countries without any data using the `geom_sf_pattern()` function from the {ggpattern} package (FC et al., 2022).

```
ggplot() +
  geom_sf(
    data = covid_sf,
    aes(fill = p, geometry = geometry),
    color = "darkgray"
  ) +
  scale_fill_gradientn(
    guide = "colorbar",
    name = "Percent",
    labels = scales::comma,
    colors = c("#BFD7EA", "#087e8b", "#0B3954"),
    na.value = NA
  ) +
  geom_sf_pattern(
    data = filter(covid_sf, is.na(p)),
    pattern = "crosshatch",
    pattern_fill = "lightgray",
    pattern_color = "lightgray",
    fill = NA,
    color = "darkgray"
```

```
) +
theme_minimal()
```

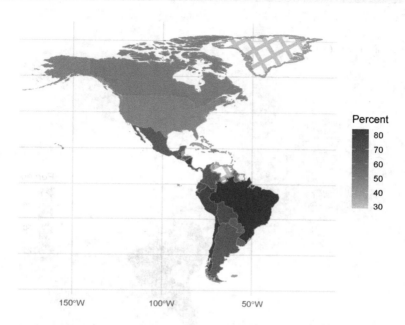

FIGURE 14.2 Percentage of households by country worried someone in their household will get COVID-19 in the next 3 months

```
ggplot() +
  geom_sf(
    data = covid_sf,
    aes(fill = p_mediumchange, geometry = geometry),
    color = "darkgray"
  ) +
  scale_fill_gradientn(
    guide = "colorbar",
    name = "Percent",
    labels = scales::comma,
    colors = c("#BFD7EA", "#087e8b", "#0B3954"),
    na.value = NA
  ) +
  geom_sf_pattern(
    data = filter(covid_sf, is.na(p_mediumchange)),
    pattern = "crosshatch",
```

```
    pattern_fill = "lightgray",
    pattern_color = "lightgray",
    fill = NA,
    color = "darkgray"
  ) +
theme_minimal()
```

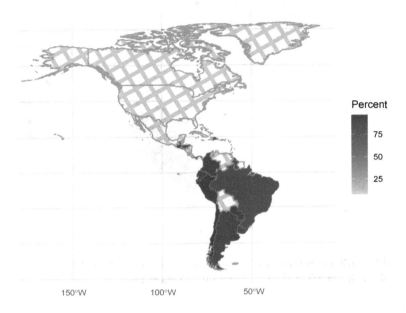

FIGURE 14.3 Percentage of households by country who had at least one child participate in virtual or hybrid learning

In Figure 14.3, we observe missing data (represented by the crosshatch pattern) for Canada, Mexico, and the United States. The questionnaires indicate that these three countries did not include the education question in the survey. To focus on countries with available data, we can remove North America from the map and show only Central and South America. We do this below by restricting the shape files to Latin America and the Caribbean, as depicted in Figure 14.4.

```
covid_c_s <- covid_sf %>%
  filter(region_wb == "Latin America & Caribbean")

ggplot() +
  geom_sf(
```

```
  data = covid_c_s,
  aes(fill = p_mediumchange, geometry = geometry),
  color = "darkgray"
) +
scale_fill_gradientn(
  guide = "colorbar",
  name = "Percent",
  labels = scales::comma,
  colors = c("#BFD7EA", "#087e8b", "#0B3954"),
  na.value = NA
) +
geom_sf_pattern(
  data = filter(covid_c_s, is.na(p_mediumchange)),
  pattern = "crosshatch",
  pattern_fill = "lightgray",
  pattern_color = "lightgray",
  fill = NA,
  color = "darkgray"
) +
theme_minimal()
```

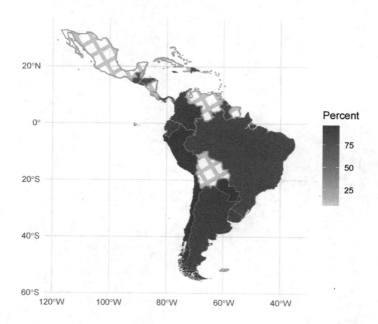

FIGURE 14.4 Percentage of households who had at least one child participate in virtual or hybrid learning, in Central and South America

In Figure 14.4, we can see that most countries with available data have similar percentages (reflected in their similar shades). However, Haiti stands out with a lighter shade, indicating a considerably lower percentage of households with at least one child participating in virtual or hybrid learning.

14.7   Exercises

1. Calculate the percentage of households with broadband internet and those with any internet at home, including from a phone or tablet in Latin America and the Caribbean. Hint: if there are countries with 0% internet usage, try filtering by something first.

2. Create a faceted map showing both broadband internet and any internet usage.

A

Importing survey data into R

To analyze a survey, we need to bring the survey data into R. This process is often referred to as importing, loading, or reading in data. Survey files come in different formats depending on the software used to create them. One of the many advantages of R is its flexibility in handling various data formats, regardless of their file extensions. Here are examples of common public-use survey file formats we may encounter:

- Delimiter-separated text files
- Excel spreadsheets in .xls or .xlsx format
- R native .rda files
- Stata datasets in .dta format
- SAS datasets in .sas format
- SPSS datasets in .sav format
- Application Programming Interfaces (APIs), often in JavaScript Object Notation (JSON) format
- Data stored in databases

This appendix guides analysts through the process of importing these various types of survey data into R.

A.1 Importing delimiter-separated files into R

Delimiter-separated files use specific characters, known as delimiters, to separate values within the file. For example, CSV (comma-separated values) files use commas as delimiters, while TSV (tab-separated values) files use tabs. These file formats are widely used because of their simplicity and compatibility with various software applications.

The {readr} package, part of the tidyverse ecosystem, offers efficient ways to import delimiter-separated files into R (Wickham et al., 2024). It offers several advantages, including automatic data type detection and flexible handling of missing values, depending on one's survey analysis needs. The {readr} package includes functions for:

- `read_csv()`: This function is specifically designed to read CSV files.
- `read_tsv()`: Use this function for TSV files.
- `read_delim()`: This function can handle a broader range of delimiter-separated files, including CSV and TSV. Specify the delimiter using the `delim` argument.
- `read_fwf()`: This function is useful for importing fixed-width files (FWF), where columns have predetermined widths, and values are aligned in specific positions.
- `read_table()`: Use this function when dealing with whitespace-separated files, such as those with spaces or multiple spaces as delimiters.
- `read_log()`: This function can read and parse web log files.

The syntax for `read_csv()` is:

```
read_csv(
  file,
  col_names = TRUE,
  col_types = NULL,
  col_select = NULL,
  id = NULL,
  locale = default_locale(),
  na = c("", "NA"),
  comment = "",
  trim_ws = TRUE,
  skip = 0,
  n_max = Inf,
  guess_max = min(1000, n_max),
  name_repair = "unique",
  num_threads = readr_threads(),
  progress = show_progress(),
  show_col_types = should_show_types(),
  skip_empty_rows = TRUE,
  lazy = should_read_lazy()
)
```

The arguments are:

- `file`: the path to the CSV file to import
- `col_names`: a value of `TRUE` imports the first row of the `file` as column names and not included in the data frame. A value of `FALSE` creates automated column names. Alternatively, we can provide a vector of column names.
- `col_types`: by default, R infers the column variable types. We can also provide a column specification using `list()` or `cols()`; for example, use `col_types = cols(.default = "c")` to read all the columns as characters. Alternatively, we can use a string to specify the variable types for each column.
- `col_select`: the columns to include in the results

- id: a column for storing the file path. This is useful for keeping track of the input file when importing multiple CSVs at a time.
- locale: the location-specific defaults for the file
- na: a character vector of values to interpret as missing
- comment: a character vector of values to interpret as comments
- trim_ws: a value of TRUE trims leading and trailing white space
- skip: number of lines to skip before importing the data
- n_max: maximum number of lines to read
- guess_max: maximum number of lines used for guessing column types
- name_repair: whether to check column names. By default, the column names are unique.
- num_threads: the number of processing threads to use for initial parsing and lazy reading of data
- progress: a value of TRUE displays a progress bar
- show_col_types: a value of TRUE displays the column types
- skip_empty_rows: a value of TRUE ignores blank rows
- lazy: a value of TRUE reads values lazily

The other functions share a similar syntax to read_csv(). To find more details, run ?? followed by the function name. For example, run ??read_tsv in the Console for additional information on importing TSV files.

In the example below, we use {readr} to import a CSV file named 'anes_timeseries_2020_csv_20220210.csv' into an R object called anes_csv. The read_csv() imports the file and stores the data in the anes_csv object. We can then use this object for further analysis.

```
library(readr)

anes_csv <-
  read_csv(file = "data/anes_timeseries_2020_csv_20220210.csv")
```

A.2 Importing Excel files into R

Excel, a widely used spreadsheet software program created by Microsoft, is a common file format in survey research. We can import Excel spreadsheets into the R environment using the {readxl} package. The package supports both the legacy .xls files and the modern .xlsx format.

To import Excel data into R, we can use the read_excel() function from the {readxl} package. This function offers a range of options for the import process. Let's explore the syntax:

```
read_excel(
  path,
  sheet = NULL,
  range = NULL,
  col_names = TRUE,
  col_types = NULL,
  na = "",
  trim_ws = TRUE,
  skip = 0,
  n_max = Inf,
  guess_max = min(1000, n_max),
  progress = readxl_progress(),
  .name_repair = "unique"
)
```

The arguments are:

- path: the path to the Excel file to import
- sheet: the name or index of the sheet (sometimes called tabs) within the Excel file
- range: the range of cells to import (for example, P15:T87)
- col_names: indicates whether the first row of the dataset contains column names
- col_types: specifies the data types of columns
- na: defines the representation of missing values (for example, NULL)
- trim_ws: controls whether leading and trailing whitespaces should be trimmed
- skip and n_max: enable skipping rows and limit the number of rows imported
- guess_max: sets the maximum number of rows used for data type guessing
- progress: specifies a progress bar for large imports
- .name_repair: determines how column names are repaired if they are not valid

In the code example below, we import an Excel spreadsheet named 'anes_timeseries_2020_csv_20220210.xlsx' into R. The resulting data is saved as a tibble in the anes_excel object, ready for further analysis.

```
library(readxl)

anes_excel <-
  read_excel(path = "data/anes_timeseries_2020_csv_20220210.xlsx")
```

A.3 Importing Stata, SAS, and SPSS files into R

The {haven} package, also from the tidyverse ecosystem, imports various proprietary data formats: Stata .dta files, SPSS .sav files, and SAS .sas7bdat and .sas7bcat files (Wickham et al., 2023b). One of the notable strengths of the {haven} package is its ability to handle multiple proprietary formats within a unified framework. It offers dedicated functions for each supported proprietary format, making it straightforward to import data regardless of the program. Here, we introduce read_dat() for Stata files, read_sav() for SPSS files, and read_sas() for SAS files.

A.3.1 Syntax

Let's explore the syntax for importing Stata files .dat files using haven::read_dat():

```
read_dta(
  file,
  encoding = NULL,
  col_select = NULL,
  skip = 0,
  n_max = Inf,
  .name_repair = "unique"
)
```

The arguments are:

- file: the path to the proprietary data file to import
- encoding: specifies the character encoding of the data file
- col_select: selects specific columns for import
- skip and n_max: control the number of rows skipped and the maximum number of rows imported
- .name_repair: determines how column names are repaired if they are not valid

The syntax for read_sav() is similar to read_dat():

```
read_sav(
  file,
  encoding = NULL,
  user_na = FALSE,
  col_select = NULL,
```

```
    skip = 0,
    n_max = Inf,
    .name_repair = "unique"
)
```

The arguments are:

- `file`: the path to the proprietary data file to import
- `encoding`: specifies the character encoding of the data file
- `col_select`: selects specific columns for import
- `user_na`: a value of `TRUE` reads variables with user-defined missing labels into `labelled_spss()` objects
- `skip` and `n_max`: control the number of rows skipped and the maximum number of rows imported
- `.name_repair`: determines how column names are repaired if they are not valid

The syntax for importing SAS files with `read_sas()` is as follows:

```
read_sas(
    data_file,
    catalog_file = NULL,
    encoding = NULL,
    catalog_encoding = encoding,
    col_select = NULL,
    skip = 0L,
    n_max = Inf,
    .name_repair = "unique"
)
```

The arguments are:

- `data_file`: the path to the proprietary data file to import
- `catalog_file`: the path to the catalog file to import
- `encoding`: specifies the character encoding of the data file
- `catalog_encoding`: specifies the character encoding of the catalog file
- `col_select`: selects specific columns for import
- `skip` and `n_max`: control the number of rows skipped and the maximum number of rows imported
- `.name_repair`: determines how column names are repaired if they are not valid

In the code examples below, we demonstrate how to import Stata, SPSS, and SAS files into R using the respective {haven} functions. The resulting data are stored in `anes_dta`, `anes_sav`, and `anes_sas` objects as tibbles, ready for use in R. For the Stata example, we show how to import the data from the {srvyrexploR} package to use in examples.

Stata:

```
library(haven)

anes_dta <-
  read_dta(file = system.file("extdata",
    "anes_2020_stata_example.dta",
    package = "srvyrexploR"
  ))
```

SPSS:

```
library(haven)

anes_sav <-
  read_sav(file = "data/anes_timeseries_2020_spss_20220210.sav")
```

SAS:

```
library(haven)

anes_sas <-
  read_sas(
    data_file = "data/anes_timeseries_2020_sas_20220210.sas7bdat"
  )
```

A.3.2 Working with labeled data

Stata, SPSS, and SAS files can contain labeled variables and values. These labels provide descriptive information about categorical data, making them easier to understand and analyze. When importing data from Stata, SPSS, or SAS, we want to preserve these labels to maintain data fidelity.

Consider a variable like 'Education Level' with coded values (e.g., 1, 2, 3). Without labels, these codes can be cryptic. However, with labels ('High School Graduate,' 'Bachelor's Degree,' 'Master's Degree'), the data become more informative and easier to work with.

With the {haven} package, we have the capability to import and work with labeled data from Stata, SPSS, and SAS files. The package uses a special class of data called `haven_labelled` to store labeled variables. When a dataset label is defined in Stata, it is stored in the 'label' attribute of the tibble when imported, ensuring that the information is not lost.

We can use functions like `select()`, `glimpse()`, and `is.labelled()` to inspect the imported data and verify if the variables are labeled. Take a look at the ANES Stata file. Notice that categorical variables V200002 and V201006 are marked with a type of `<dbl+lbl>`. This notation indicates that these variables are labeled.

```
library(dplyr)

anes_dta %>%
  select(1:6) %>%
  glimpse()
```

```
## Rows: 7,453
## Columns: 6
## $ V200001  <dbl> 200015, 200022, 200039, 200046, 200053, 200060, 200~
## $ V200002  <dbl+lbl> 3, 3, 3, 3, 3, 3, 3, 3, 3, 3, 3, 3, 3, 3, 3, 3,~
## $ V200010b <dbl> 1.0057, 1.1635, 0.7687, 0.5210, 0.9658, 0.2347, 0.4~
## $ V200010d <dbl> 9, 26, 41, 29, 23, 37, 7, 37, 32, 41, 22, 7, 38, 21~
## $ V200010c <dbl> 2, 2, 1, 2, 1, 2, 1, 2, 2, 2, 1, 1, 2, 2, 2, 2, 1, ~
## $ V201006  <dbl+lbl> 2, 3, 2, 3, 2, 1, 2, 3, 2, 2, 2, 2, 2, 1, 2, 1,~
```

We can confirm their label status using the `haven::is.labelled()` function.

```
haven::is.labelled(anes_dta$V200002)
```

```
## [1] TRUE
```

To explore the labels further, we can use the `attributes()` function. This function provides insights into both the variable labels (`$label`) and the associated value labels (`$labels`).

```
attributes(anes_dta$V200002)
```

```
## $label
## [1] "Mode of interview: pre-election interview"
##
## $format.stata
## [1] "%10.0g"
##
## $class
## [1] "haven_labelled" "vctrs_vctr"      "double"
##
## $labels
##     1. Video 2. Telephone      3. Web
##             1             2           3
```

When we import a labeled dataset using {haven}, it results in a tibble containing both the data and label information. However, this is meant to be an intermediary data structure and not intended to be the final data format for analysis. Instead, we should convert it into a regular R data frame before continuing our data workflow. There are two primary methods to achieve this conversion: (1) convert to factors or (2) remove the labels.

Option 1: Convert the vector into a factor

Factors are native R data types for working with categorical data. They consist of integer values that correspond to character values, known as levels. Below is a dummy example of factors. The `factors` show the four different levels in the data: `strongly agree, agree, disagree,` and `strongly disagree`.

```
response <-
  c("strongly agree", "agree", "agree", "disagree", "strongly disagree")

response_levels <-
  c("strongly agree", "agree", "disagree", "strongly disagree")

factors <- factor(response, levels = response_levels)

factors
```

```
## [1] strongly agree    agree              agree
## [4] disagree          strongly disagree
## Levels: strongly agree agree disagree strongly disagree
```

Factors are integer vectors, though they may look like character strings. We can confirm by looking at the vector's structure:

```
glimpse(factors)
```

```
##  Factor w/ 4 levels "strongly agree",..: 1 2 2 3 4
```

R's factors differ from Stata, SPSS, or SAS labeled vectors. However, we can convert labeled variables into factors using the `as_factor()` function.

```
anes_dta %>%
  transmute(V200002 = as_factor(V200002))
```

```
## # A tibble: 7,453 x 1
##    V200002
##    <fct>
## 1 3. Web
## 2 3. Web
```

```
##  3 3. Web
##  4 3. Web
##  5 3. Web
##  6 3. Web
##  7 3. Web
##  8 3. Web
##  9 3. Web
## 10 3. Web
## # i 7,443 more rows
```

The `as_factor()` function can be applied to all columns in a data frame or individual ones. Below, we convert all `<dbl+lbl>` columns into factors.

```
anes_dta_factor <-
  anes_dta %>%
  as_factor()

anes_dta_factor %>%
  select(1:6) %>%
  glimpse()
```

```
## Rows: 7,453
## Columns: 6
## $ V200001  <dbl> 200015, 200022, 200039, 200046, 200053, 200060, 200~
## $ V200002  <fct> 3. Web, 3. Web, 3. Web, 3. Web, 3. Web, 3. Web, 3. ~
## $ V200010b <dbl> 1.0057, 1.1635, 0.7687, 0.5210, 0.9658, 0.2347, 0.4~
## $ V200010d <dbl> 9, 26, 41, 29, 23, 37, 7, 37, 32, 41, 22, 7, 38, 21~
## $ V200010c <dbl> 2, 2, 1, 2, 1, 2, 1, 2, 2, 1, 1, 2, 2, 2, 2, 1, ~
## $ V201006  <fct> 2. Somewhat interested, 3. Not much interested, 2. ~
```

Option 2: Strip the labels

The second option is to remove the labels altogether, converting the labeled data into a regular R data frame. To remove, or 'zap,' the labels from our tibble, we can use the {haven} package's `zap_label()` and `zap_labels()` functions. This approach removes the labels but retains the data values in their original form.

The ANES Stata file columns contain variable labels. Using the `map()` function from {purrr}, we can review the labels using `attr`. In the example below, we list the first two variables and their labels. For instance, the label for `V200002` is "Mode of interview: pre-election interview."

```
purrr::map(anes_dta, ~ attr(.x, "label")) %>%
  head(2)
```

```
## $V200001
## [1] "2020 Case ID"
##
## $V200002
## [1] "Mode of interview: pre-election interview"
```

Use `zap_label()` to remove the variable labels but retain the value labels. Notice that the labels return as NULL.

```
zap_label(anes_dta) %>%
  purrr::map(~ attr(.x, "label")) %>%
  head(2)
```

```
## $V200001
## NULL
##
## $V200002
##       1. Video 2. Telephone       3. Web
##              1            2            3
```

To remove the value labels, use `zap_labels()`. Notice the previous <dbl+lbl> columns are now <dbl>.

```
zap_labels(anes_dta) %>%
  select(1:6) %>%
  glimpse()
```

```
## Rows: 7,453
## Columns: 6
## $ V200001  <dbl> 200015, 200022, 200039, 200046, 200053, 200060, 200~
## $ V200002  <dbl> 3, 3, 3, 3, 3, 3, 3, 3, 3, 3, 3, 3, 3, 3, 3, 3, ~
## $ V200010b <dbl> 1.0057, 1.1635, 0.7687, 0.5210, 0.9658, 0.2347, 0.4~
## $ V200010d <dbl> 9, 26, 41, 29, 23, 37, 7, 37, 32, 41, 22, 7, 38, 21~
## $ V200010c <dbl> 2, 2, 1, 2, 1, 2, 1, 2, 2, 2, 1, 1, 2, 2, 2, 2, 1, ~
## $ V201006  <dbl> 2, 3, 2, 3, 2, 1, 2, 3, 2, 2, 2, 2, 2, 1, 2, 1, 1, ~
```

While it is important to convert labeled datasets into regular R data frames for working in R, the labels themselves often contain valuable information that provides context and meaning to the survey variables. To aid with interpretability and documentation, we can create a data dictionary from the labeled dataset. A data dictionary is a reference document that provides detailed information about the variables and values of a survey.

The {labelled} package offers a convenient function, `generate_dictionary()`, that creates data dictionaries directly from a labeled dataset (Larmarange, 2024). This function extracts variable labels, value labels, and other metadata and organizes them into a structured document that we can browse and reference throughout our analysis.

Let's create a data dictionary from the ANES Stata dataset as an example:

```
library(labelled)

dictionary <- generate_dictionary(anes_dta)
```

Once we've generated the data dictionary, we can take a look at the `V200002` variable and see the label, column type, number of missing entries, and associated values.

```
dictionary %>%
  filter(variable == "V200002")
```

```
## pos variable label              col_type missing values
## 2   V200002  Mode of interview: p~ dbl+lbl  0       [1] 1. Video
##                                                      [2] 2. Telephone
##                                                      [3] 3. Web
```

A.3.3 Labeled missing data values

In survey data analysis, dealing with missing values is a crucial aspect of data preparation. Stata, SPSS, and SAS files each have their own method for handling missing values.

- Stata has "extended" missing values, `.A` through `.Z`.
- SAS has "special" missing values, `.A` through `.Z` and `._`.
- SPSS has per-column "user" missing values. Each column can declare up to three distinct values or a range of values (plus one distinct value) that should be treated as missing.

SAS and Stata use a concept known as 'tagged' missing values, which extend R's regular `NA`. A 'tagged' missing value is essentially an `NA` with an additional single-character label. These values behave identically to regular `NA` in standard R operations while preserving the informative tag associated with the missing value.

Here is an example from the NORC at the University of Chicago's 2018 General Society Survey, where Don't Know (`DK`) responses are tagged as `NA(d)`, Inapplicable (`IAP`) responses are tagged as `NA(i)`, and `No Answer` responses are tagged as `NA(n)` (Davern et al., 2021).

```
head(gss_dta$HEALTH)
#> <labelled<double>[6]>: condition of health
#> [1]      2      1 NA(i) NA(i)      1      2
#>
#> Labels:
#>  value     label
#>      1 excellent
#>      2      good
#>      3      fair
#>      4      poor
#>  NA(d)        DK
#>  NA(i)       IAP
#>  NA(n)        NA
```

In contrast, SPSS uses a different approach called 'user-defined values' to
denote missing values. Each column in an SPSS dataset can have up to
three distinct values designated as missing or a specified range of missing
values. To model these additional user-defined missing values, {haven} provides
the labeled_spss() subclass of labeled(). When importing SPSS data using
{haven}, it ensures that user-defined missing values are correctly handled. We
can work with these data in R while preserving the unique missing value
conventions from SPSS.

Here is what the GSS SPSS dataset looks like when loaded with {haven}.

```
head(gss_sps$HEALTH)
#> <labelled_spss<double>[6]>: Condition of health
#> [1] 2 1 0 0 1 2
#> Missing values: 0, 8, 9
#>
#> Labels:
#>  value     label
#>      0       IAP
#>      1 EXCELLENT
#>      2      GOOD
#>      3      FAIR
#>      4      POOR
#>      8        DK
#>      9        NA
```

A.4   Importing data from APIs into R

In addition to working with data saved as files, we may also need to retrieve data through Application Programming Interfaces (APIs). APIs provide a structured way to access data hosted on external servers and import them directly into R for analysis.

To access these data, we need to understand how to construct API requests. Each API has unique endpoints, parameters, and authentication requirements. Pay attention to:

- Endpoints: These are URLs that point to specific data or services
- Parameters: Information passed to the API to customize the request (e.g., date ranges, filters)
- Authentication: APIs may require API keys or tokens for access
- Rate Limits: APIs may have usage limits, so be aware of any rate limits or quotas

Typically, we begin by making a GET request to an API endpoint. The {httr2} package allows us to generate and process HTTP requests (Wickham, 2024). We can make the GET request by pointing to the URL that contains the data we would like:

```r
library(httr2)

api_url <- "https://api.example.com/survey-data"
response <- GET(url = api_url)
```

Once we make the request, we obtain the data as the `response`. The data often come in JSON format. We can extract and parse the data using the {jsonlite} package, allowing us to work with them in R (Ooms, 2014). The `fromJSON()` function, shown below, converts JSON data to an R object.

```r
survey_data <- fromJSON(content(response, "text"))
```

Note that these are dummy examples. Please review the documentation to understand how to make requests from a specific API.

R offers several packages that simplify API access by providing ready-to-use functions for popular APIs. These packages are called "wrappers," as they "wrap" the API in R to make it easier to use. For example, the {tidycensus} package used in this book simplifies access to U.S. Census data, allowing us to retrieve data with R commands instead of writing API requests from scratch (Walker and Herman, 2024). Behind the scenes, `get_pums()` is making a GET request from the Census API, and the {tidycensus} functions are converting the response into an R-friendly format. For example, if we are interested in the age, sex, race, and Hispanicity of those in the American Community Survey sample of Durham County, North Carolina[1], we can use the get_pums() function to extract the microdata as shown in the code below. We can then use the replicate weights to create a survey object and calculate estimates for Durham County.

```
library(tidycensus)

durh_pums <- get_pums(
  variables = c("PUMA", "SEX", "AGEP", "RAC1P", "HISP"),
  state = "NC",
  puma = c("01301", "01302"),
  survey = "acs1",
  year = 2022,
  rep_weights = "person"
)
```

```
## Getting data from the 2022 1-year ACS Public Use Microdata Sample
```

```
durh_pums
```

```
## # A tibble: 2,724 x 90
##    SERIALNO   SPORDER AGEP PUMA  ST    SEX   HISP  RAC1P WGTP PWGTP
##    <chr>        <dbl> <dbl> <chr> <chr> <chr> <chr> <chr> <dbl> <dbl>
##  1 2022HU0427~      1    69 01301 37    2     01    1      104   104
##  2 2022HU0431~      1    62 01301 37    2     01    2      145   145
##  3 2022HU0432~      1    48 01301 37    1     01    6       63    63
##  4 2022HU0432~      1    70 01301 37    2     01    1      102   101
##  5 2022HU0432~      2    69 01301 37    1     01    1      102    74
##  6 2022HU0432~      1    74 01302 37    2     01    1      123   124
##  7 2022HU0432~      1    26 01301 37    2     01    6       22    23
##  8 2022HU0432~      2    35 01301 37    2     01    2       22    65
##  9 2022HU0432~      3    28 01301 37    2     01    6       22    25
## 10 2022HU0432~      4    25 01301 37    2     01    1       22    48
## # i 2,714 more rows
## # i 80 more variables: PWGTP1 <dbl>, PWGTP2 <dbl>, PWGTP3 <dbl>,
## #   PWGTP4 <dbl>, PWGTP5 <dbl>, PWGTP6 <dbl>, PWGTP7 <dbl>,
## #   PWGTP8 <dbl>, PWGTP9 <dbl>, PWGTP10 <dbl>, PWGTP11 <dbl>,
## #   PWGTP12 <dbl>, PWGTP13 <dbl>, PWGTP14 <dbl>, PWGTP15 <dbl>,
## #   PWGTP16 <dbl>, PWGTP17 <dbl>, PWGTP18 <dbl>, PWGTP19 <dbl>,
## #   PWGTP20 <dbl>, PWGTP21 <dbl>, PWGTP22 <dbl>, PWGTP23 <dbl>, ...
```

[1]The public use microdata areas (PUMA) for Durham County were identified using the 2020 PUMA Names File: https://www2.census.gov/geo/pdfs/reference/puma2020/2020_PUMA_Names.pdf

In Chapter 4, we used the {censusapi} package to get data from the Census data API for the Current Population Survey. To discover if there is an R package that directly interfaces with a specific survey or data source, search for "[survey] R wrapper" or "[data source] R package" online.

A.5 Importing data from databases in R

Databases provide a secure and organized solution as the volume and complexity of data grow. We can access, manage, and update data stored in databases in a systematic way. Because of how the data are organized, teams can draw from the same source and obtain any metadata that would be helpful for analysis.

There are various ways of using R to work with databases. If using RStudio, we can connect to different databases through the Connections Pane in the top right of the IDE. We can also use packages like {DBI} and {odbc} to access database tables in R files. Here is an example script connecting to a database:

```
con <-
  DBI::dbConnect(
    odbc::odbc(),
    Driver    = "[driver name]",
    Server    = "[server path]",
    UID       = rstudioapi::askForPassword("Database user"),
    PWD       = rstudioapi::askForPassword("Database password"),
    Database  = "[database name]",
    Warehouse = "[warehouse name]",
    Schema    = "[schema name]"
  )
```

The {dbplyr} and {dplyr} packages allow us to make queries and run data analysis entirely using {dplyr} syntax. All of the code can be written in R, so we do not have to switch between R and SQL to explore the data. Here is some sample code:

```
q1 <- tbl(con, "bank") %>%
  group_by(month_idx, year, month) %>%
  summarize(subscribe = sum(ifelse(term_deposit == "yes", 1, 0)),
            total = n())

show_query(q1)
```

Be sure to check the documentation to configure a database connection.

A.6 Importing data from other formats

R also offers dedicated packages such as {googlesheets4} for Google Sheets or {qualtRics} for Qualtrics. With less common or proprietary file formats, the broader data science community can often provide guidance. Online resources like Stack Overflow[2] and dedicated forums like Posit Community[3] are valuable sources of information for importing data into R.

---

[2] https://stackoverflow.com/
[3] https://forum.posit.co/

B

ANES derived variable codebook

The full codebook with the original variables is available at American National Election Studies (2022).

The ANES codebook for the data used in this book (`anes_2020` from {srvyrexploR}) is available in the online version of the book at https://tidy-survey-r.github.io/tidy-survey-book/anes-cb.html.

C

RECS derived variable codebook

The full codebook with the original variables is available at https://www.eia.gov/consumption/residential/data/2020/index.php?view=microdata - "Variable and Response Codebook."

The RECS codebook for the data used in this book (`recs_2020` from {srvyrexploR}) is available in the online version of the book at https://tidy-survey-r.github.io/tidy-survey-book/recs-cb.html.

D

Exercise solutions

Exercise solutions are available in the online version of the book at https://tidy-survey-r.github.io/tidy-survey-book/exercise-solutions.html.

Bibliography

Allaire, J., Teague, C., Scheidegger, C., Xie, Y., and Dervieux, C. (2024). Quarto. https://github.com/quarto-dev/quarto-cli. Type: Software.

American National Election Studies (2021). Anes 2020 time series study: Pre-election and post-election survey questionnaires. https://electionstudies.org/wp-content/uploads/2021/07/anes_timeseries_2020_questionnaire_20210719.pdf.

American National Election Studies (2022). Anes 2020 time series study full release: User guide and codebook. https://electionstudies.org/wp-content/uploads/2022/02/anes_timeseries_2020_userguidecodebook_20220210.pdf.

Bache, S. M. and Wickham, H. (2022). *magrittr: A Forward-Pipe Operator for R.* R package version 2.0.3.

Biemer, P. P. (2010). Total survey error: Design, implementation, and evaluation. *Public Opinion Quarterly*, 74(5):817–848.

Biemer, P. P. and Lyberg, L. E. (2003). *Introduction to survey quality.* John Wiley & Sons.

Biemer, P. P., Murphy, J., Zimmer, S., Berry, C., Deng, G., and Lewis, K. (2017). Using Bonus Monetary Incentives to Encourage Web Response in Mixed-Mode Household Surveys. *Journal of Survey Statistics and Methodology*, 6(2):240–261.

Bollen, K. A., Biemer, P. P., Karr, A. F., Tueller, S., and Berzofsky, M. E. (2016). Are survey weights needed? a review of diagnostic tests in regression analysis. *Annual Review of Statistics and Its Application*, 3(1):375–392.

Bradburn, N. M., Sudman, S., and Wansink, B. (2004). *Asking Questions: The Definitive Guide to Questionnaire Design.* Jossey-Bass, 2nd edition.

Bryan, J. (2023). Happy Git and Github for the useR. https://happygitwithr.com/.

Centers for Disease Control and Prevention (CDC) (2021). Behavioral risk factor surveillance system survey questionnaire. https://www.cdc.gov/brfss/questionnaires/pdf-ques/2021-BRFSS-Questionnaire-1-19-2022-508.pdf.

Cochran, W. G. (1977). *Sampling techniques.* John Wiley & Sons.

Cox, B. G., Binder, D. A., Chinnappa, B. N., Christianson, A., Colledge, M. J., and Kott, P. S. (2011). *Business survey methods*. John Wiley & Sons.

Csardi, G. (2023). *prettyunits: Pretty, Human Readable Formatting of Quantities*. R package version 1.2.0.

Csárdi, G. and Hester, J. (2024). *pak: Another Approach to Package Installation*. R package version 0.7.2.

Davern, M., Bautista, R., Freese, J., Morgan, S. L., and Smith, T. W. (2021). General Social Survey 2016-2020 Panel Codebook. https://gss.norc.org /Documents/codebook/2016-2020%20GSS%20Panel%20Codebook%20- %20R1a.pdf.

DeBell, M. (2010). How to analyze anes survey data. ANES Technical Report Series nes012492, Palo Alto, CA: Stanford University and Ann Arbor, MI: the University of Michigan.

DeBell, M., Amsbary, M., Brader, T., Brock, S., Good, C., Kamens, J., Maisel, N., and Pinto, S. (2022). Methodology Report for the ANES 2020 Time Series Study. https://electionstudies.org/wp-content/uploads/2022/08/an es_timeseries_2020_methodology_report.pdf.

DeLeeuw, E. D. (2005). To mix or not to mix data collection modes in surveys. *Journal of Official Statistics*, 21:233–255.

DeLeeuw, E. D. (2018). Mixed-mode: Past, present, and future. *Survey Research Methods*, 12(2):75–89.

Deming, W. E. (1991). *Sample design in business research*, volume 23. John Wiley & Sons.

Dillman, D. A., Smyth, J. D., and Christian, L. M. (2014). *Internet, phone, mail, and mixed-mode surveys: The tailored design method*. John Wiley & Sons.

FC, M., Davis, T. L., and ggplot2 authors (2022). *ggpattern: 'ggplot2' Pattern Geoms*. R package version 1.0.1.

Fowler, F. J. and Mangione, T. W. (1989). *Standardized Survey Interviewing*. SAGE.

Freedman Ellis, G. and Schneider, B. (2024). *srvyr: 'dplyr'-Like Syntax for Summary Statistics of Survey Data*. R package version 1.3.0.

Fuller, W. A. (2011). *Sampling statistics*. John Wiley & Sons.

Gard, A. M., Hyde, L. W., Heeringa, S. G., West, B. T., and Mitchell, C. (2023). Why weight? analytic approaches for large-scale population neuroscience data. *Developmental Cognitive Neuroscience*, 59:101196.

Gelman, A. (2007). Struggles with Survey Weighting and Regression Modeling. *Statistical Science*, 22(2):153–164.

Groves, R. M., Fowler Jr, F. J., Couper, M. P., Lepkowski, J. M., Singer, E., and Tourangeau, R. (2009). *Survey methodology*. John Wiley & Sons.

Harter, R., Battaglia, M. P., Buskirk, T. D., Dillman, D. A., English, N., Fahimi, M., Frankel, M. R., Kennel, T., McMichael, J. P., McPhee, C. B., Montaquila, J., Yancey, T., and Zuckerberg, A. L. (2016). Address-based sampling. Task force report, American Association for Public Opinion Research.

Henry, L. and Wickham, H. (2024). *tidyselect: Select from a Set of Strings*. R package version 1.2.1.

Iannone, R., Cheng, J., Schloerke, B., Hughes, E., Lauer, A., Seo, J., Brevoort, K., and Roy, O. (2024). *gt: Easily Create Presentation-Ready Display Tables*. R package version 0.11.0.9000, commit 28de628ac53ecded0f747396333507477f537923.

Kim, J. K. and Shao, J. (2021). *Statistical Methods for Handling Incomplete Data*. Chapman & Hall/CRC Press.

Landau, W. M. (2021). The targets R package: a dynamic Make-like function-oriented pipeline toolkit for reproducibility and high-performance computing. *Journal of Open Source Software*, 6(57):2959.

LAPOP (2021a). AmericasBarometer 2021 - Canada: Technical information. Technical report, Vanderbilt University.

LAPOP (2021b). AmericasBarometer 2021 - U.S.: Technical information. Technical report, Vanderbilt University.

LAPOP (2021c). AmericasBarometer 2021: Technical information. Technical report, Vanderbilt University.

LAPOP (2021d). Core questionnaire. https://www.vanderbilt.edu/lapop/ab2021/AB2021-Core-Questionnaire-v17.5-Eng-210514-W-v2.pdf.

LAPOP (2023a). About the AmericasBarometer. https://www.vanderbilt.edu/lapop/about-americasbarometer.php.

LAPOP (2023b). The AmericasBarometer by the LAPOP lab. www.vanderbilt.edu/lapop.

Larmarange, J. (2024). *labelled: Manipulating Labelled Data*. R package version 2.13.0.

Levy, P. S. and Lemeshow, S. (2013). *Sampling of populations: methods and applications*. John Wiley & Sons.

Lumley, T. (2010). *Complex surveys: a guide to analysis Using R.* John Wiley & Sons.

Mack, C., Su, Z., and Westreich, D. (2018). *Types of Missing Data.* Rockville (MD): Agency for Healthcare Research and Quality (US).

Massicotte, P. and South, A. (2023). *rnaturalearth: World Map Data from Natural Earth.* R package version 1.0.1.

McCullagh, P. and Nelder, J. A. (1989). Binary data. In *Generalized linear models*, pages 98–148. Springer.

Müller, K. (2020). *here: A Simpler Way to Find Your Files.* R package version 1.0.1.

National Center for Health Statistics (2023). National Health Interview Survey, 2022 survey description. https://ftp.cdc.gov/pub/Health_Statistics/NCHS/Dataset_Documentation/NHIS/2022/srvydesc-508.pdf.

Ooms, J. (2014). The jsonlite package: A practical and consistent mapping between json data and R objects. *arXiv:1403.2805 [stat.CO]*.

Pebesma, E. and Bivand, R. (2023). *Spatial Data Science: With applications in R.* Chapman & Hall/CRC.

Penn State (2019). Stat 506: Sampling theory and methods [online course]. https://online.stat.psu.edu/stat506/.

R Core Team (2024). *R: A Language and Environment for Statistical Computing.* R Foundation for Statistical Computing, Vienna, Austria.

Recht, H. (2024). *censusapi: Retrieve Data from the Census APIs.* R package version 0.9.0.9000, commit 74334d4d180ff79477456a777235afdde164f2cd.

Robinson, D., Hayes, A., and Couch, S. (2023). *broom: Convert Statistical Objects into Tidy Tibbles.* R package version 1.0.5.

Särndal, C.-E., Swensson, B., and Wretman, J. (2003). *Model assisted survey sampling.* Springer Science & Business Media.

Schafer, J. L. and Graham, J. W. (2002). Missing data: Our view of the state of the art. *Psychological Methods*, 7:147–177.

Schouten, B., Peytchev, A., and Wagner, J. (2018). *Adaptive Survey Design.* Chapman & Hall/CRC Press.

Scott, A. (2007). Rao-scott corrections and their impact. In *Section on Survey Research Methods*, pages 3514–3518.

Shah, B. V. and Vaish, A. K. (2006). Confidence intervals for quantile estimation from complex survey data. In *Proceedings of the Section on Survey Research Methods*.

Shook-Sa, B., Couzens, G. L., and Berzofsky, M. (2015). Users' guide to the National Crime Victimization Survey (NCVS) direct variance estimation. https://bjs.ojp.gov/sites/g/files/xyckuh236/files/media/document/ncvs _variance_user_guide_11.06.14.pdf.

Sjoberg, D. D., Whiting, K., Curry, M., Lavery, J. A., and Larmarange, J. (2021). Reproducible summary tables with the gtsummary package. *The R Journal*, 13:570–580.

Skinner, C. (2009). *Chapter 15: Statistical Disclosure Control for Survey Data*, pages 381–396. Elsevier B.V.

Sprunt, B. (2020). 93 million and counting: Americans are shattering early voting records. *National Public Radio*.

Tierney, N. (2017). visdat: Visualising whole data frames. *Journal of Open Source Software*, 2(16):355.

Tierney, N. and Cook, D. (2023). Expanding tidy data principles to facilitate missing data exploration, visualization and assessment of imputations. *Journal of Statistical Software*, 105(7):1–31.

Tourangeau, R., Couper, M. P., and Conrad, F. (2004). Spacing, position, and order: Interpretive heuristics for visual features of survey questions. *Public Opinion Quarterly*, 68:368–393.

Tourangeau, R., Rips, L. J., and Rasinski, K. (2000). *Psychology of Survey Response*. Cambridge University Press.

U. S. Bureau of Justice Statistics (2017). National Crime Victimization Survey, 2016: Technical Documentation. https://bjs.ojp.gov/sites/g/files/xyckuh2 36/files/media/document/ncvstd16.pdf.

U. S. Bureau of Justice Statistics (2020). National Crime Victimization Survey ncvs-2 crime incident report. https://bjs.ojp.gov/content/pub/pdf/ncvs20 _cir.pdf.

U. S. Bureau of Justice Statistics (2022). National Crime Victimization Survey, [United States], 2021. https://www.icpsr.umich.edu/web/NACJD/studies/ 38429/datadocumentation. Download - DS0 Study-Level Files - Codebook [PDF].

U.S. Bureau of Justice Statistics (2022). National Crime Victimization Survey, [United States], 2021. https://www.icpsr.umich.edu/web/NACJD/studies/ 38429. Type: dataset.

U.S. Census Bureau (2021). Understanding and Using the American Community Survey Public Use Microdata Sample Files What Data Users Need to Know. https://www.census.gov/content/dam/Census/library/publications /2021/acs/acs_pums_handbook_2021.pdf.

U.S. Energy Information Administration (2017). Residential Energy Consumption Survey (RECS): Using the 2015 microdata file to compute estimates and standard errors (RSEs). https://www.eia.gov/consumption/residential/data/2015/pdf/microdata_v3.pdf.

U.S. Energy Information Administration (2020). Residential energy consumption survey (recs) form eia-457a 2020 household questionnaire. https://www.eia.gov/survey/form/eia_457/archive/2020_RECS-457A.pdf.

U.S. Energy Information Administration (2023a). 2020 Residential Energy Consumption Survey: Consumption and Expenditures Technical Documentation Summary. https://www.eia.gov/consumption/residential/data/2020/pdf/2020%20RECS%20CE%20Methodology_Final.pdf.

U.S. Energy Information Administration (2023b). 2020 Residential Energy Consumption Survey: Household Characteristics Technical Documentation Summary. https://www.eia.gov/consumption/residential/data/2020/pdf/2020%20RECS_Methodology%20Report.pdf.

U.S. Energy Information Administration (2023c). 2020 Residential Energy Consumption Survey: Using the microdata file to compute estimates and relative standard errors (RSEs). https://www.eia.gov/consumption/residential/data/2020/pdf/microdata-guide.pdf.

U.S. Energy Information Administration (2023d). Units and calculators explained: Degree days. https://www.eia.gov/energyexplained/units-and-calculators/degree-days.php.

Ushey, K. and Wickham, H. (2024). *renv: Project Environments.* R package version 1.0.7.

Valliant, R. and Dever, J. A. (2018). *Survey Weights: A Step-by-step Guide to Calculation.* Stata Press.

Valliant, R., Dever, J. A., and Kreuter, F. (2013). *Practical tools for designing and weighting survey samples*, volume 1. Springer.

Walker, K. and Herman, M. (2024). *tidycensus: Load US Census Boundary and Attribute Data as 'tidyverse' and 'sf'-Ready Data Frames.* R package version 1.6.3.

Wickham, H. (2016). *ggplot2: Elegant Graphics for Data Analysis.* Springer-Verlag New York.

Wickham, H. (2019). *Advanced R.* CRC Press.

Wickham, H. (2023). *forcats: Tools for Working with Categorical Variables (Factors).* R package version 1.0.0.

Wickham, H. (2024). *httr2: Perform HTTP Requests and Process the Responses.* R package version 1.0.1.

Wickham, H., Averick, M., Bryan, J., Chang, W., McGowan, L. D., François, R., Grolemund, G., Hayes, A., Henry, L., Hester, J., Kuhn, M., Pedersen, T. L., Miller, E., Bache, S. M., Müller, K., Ooms, J., Robinson, D., Seidel, D. P., Spinu, V., Takahashi, K., Vaughan, D., Wilke, C., Woo, K., and Yutani, H. (2019). Welcome to the tidyverse. *Journal of Open Source Software*, 4(43):1686.

Wickham, H., François, R., Henry, L., Müller, K., and Vaughan, D. (2023a). *dplyr: A Grammar of Data Manipulation*. R package version 1.1.4.

Wickham, H. and Henry, L. (2023). *purrr: Functional Programming Tools*. R package version 1.0.2.

Wickham, H., Hester, J., and Bryan, J. (2024). *readr: Read Rectangular Text Data*. R package version 2.1.5.

Wickham, H., Miller, E., and Smith, D. (2023b). *haven: Import and Export 'SPSS', 'Stata' and 'SAS' Files*. R package version 2.5.4.

Wickham, H., Çetinkaya Rundel, M., and Grolemund, G. (2023c). *R for Data Science: Import, Tidy, Transform, Visualize, and Model Data*. O'Reilly Media, 2nd edition.

Wolter, K. M. (2007). *Introduction to variance estimation*, volume 53. Springer.

Xie, Y., Dervieux, C., and Riederer, E. (2020). *R Markdown Cookbook*. Chapman & Hall/CRC.

Zimmer, S., Powell, R., and Velásquez, I. (2024). *srvyrexploR: Data Supplement for Exploring Complex Survey Data Analysis Using R*. R package version 1.0.1, commit cdf93166a998c083e774a83c46703a408416380c.

Index